Urban Agriculture

Series Editors

For the Book Series Editors, the main objective of this series is to mobilize and enhance capacities to share UA experiences and research results, compare methodologies and tools, identify technological obstacles, and adapt solutions. By diffusing this knowledge, the aim is to contribute to building the capacity of policy-makers, professionals and practitioners in governments, international agencies, civil society, the private sector as well as academia, to effectively incorporate UA in their field of interests. It is also to constitute a global research community to debate the lessons from UA initiatives, to compare approaches, and to supply tools for aiding in the conception and evaluation of various strategies of UA development.

The concerned scientific field of this series is large because UA combines agricultural issues with those related to city management and development. Thus this interdisciplinary Book Series brings together environmental sciences, agronomy, urban and regional planning, architecture, landscape design, economics, social sciences, soil sciences, public health and nutrition, recognizing UA's contribution to meeting society's basic needs, feeding people, structuring the cities while shaping their development. All these scientific fields are of interest for this Book Series. Books in this Series will analyze UA research and actions; program implementation, urban policies, technological innovations, social and economic development, management of resources (soil/land, water, wastes…) for or by urban agriculture, are all pertinent here.

This Book Series includes a mix of edited, coauthored, and single-authored books. These books could be based on research programs, conference papers, or other collective efforts, as well as completed theses or entirely new manuscripts.

More information about this series at http://www.springer.com/series/11815

Isha DeCoito • Amie Patchen
Neil Knobloch • Levon Esters
Editors

Teaching and Learning in Urban Agricultural Community Contexts

Springer

Editors
Isha DeCoito
Western University
London, ON, Canada

Amie Patchen
Public Health Prog., Col. of Vet. Med.
Cornell University
Ithaca, NY, USA

Neil Knobloch
College of Agriculture, ASEC Department
Purdue University System
Lafayette, IN, USA

Levon Esters
College of Agriculture, ASEC Department
Purdue University System
West Lafayette, IN, USA

ISSN 2197-1730 ISSN 2197-1749 (electronic)
Urban Agriculture
ISBN 978-3-030-72887-8 ISBN 978-3-030-72888-5 (eBook)
https://doi.org/10.1007/978-3-030-72888-5

This Springer imprint is published by the registered company Springer Nature Switzerland AG
The registered company address is: Gewerbestrasse 11, 6330 Cham, Switzerland

Contents

Contributors

Daniela Benavides ConCiencia, NGO, Lima, Peru

NatureForAll, International Union for Conservation of Nature, Gland, Switzerland

Bertram C. Bruce School of Information Sciences, University of Illinois, Champaign, IL, USA

Josh Corlew Independent Scholar, Nashville, TN, USA

Kristian Demary The Best Bees Company Inc., Boston, MA, USA

Urban Beekeeping Laboratory and Bee Sanctuary Inc., Boston, MA, USA

Liberal Arts, Massachusetts College of Art and Design, Boston, MA, USA

Helena K. Farrell Former Lecturer and Visiting Scholar, University of Massachusetts, Amherst, MA, USA

Landscape Architecture & Regional Planning, Stockbridge School of Agriculture, Amherst, MA, USA

Zev H. S. Friedman Living Systems Design, Bainbridge Island, WA, USA

Heather Gillich Healthy Living Initiative, City of Minneapolis, Minneapolis, MN, USA

Adriana Gonzalez-Pestana ConCiencia NGO, Lima, Peru

Phyllis Katz Living Systems Design, Bainbridge Island, WA, USA

Neil Knobloch College of Agriculture, ASEC Department, Purdue University System, Lafayette, IN, USA

Joaquín Leguía Asociación para la Niñez y su Ambiente (ANIA), Magdalena del Mar, Lima, Peru

Christopher D. Murakami Chatham University, Pittsburgh, PA, USA

Thomas Schmitt Departments of Biology and Marine and Environmental Studies, Northeastern University, Boston, MA, USA

The Best Bees Company, Boston, MA, USA

Urban Beekeeping Laboratory and Bee Sanctuary Inc., Boston, MA, USA

Carolyn Schroeder College of Science, Texas A&M University, College Station, TX, USA

Timothy Scott Office of the Provost, Texas A&M University, College Station, TX, USA

Kevin Slick The Dalton School, New York, NY, USA

Mila Tewell The Dalton School, New York, NY, USA

Craig Wilson Texas A&M University, USDA Southern Plains Agricultural Research Center (SPARC), College Station, TX, USA

Noah Wilson-Rich The Best Bees Company Inc., Boston, MA, USA

Urban Beekeeping Laboratory and Bee Sanctuary Inc., Boston, MA, USA

Mihye Won School of Education, Curtin University, Perth, Australia

Michelle Wooten Astrophysical and Planetary Sciences Department, University of Colorado Boulder, Boulder, CO, USA

Chapter 1
An Introduction: Teaching and Learning in Urban Agricultural Community Contexts

Neil Knobloch

Abstract Urban agriculture has seen a resurgence in interest and activity as consumers are interested in where their food comes from, how it was produced, and how it reflects their personal values. The interest in urban agriculture has resulted in formal, nonformal and informal education programs, projects, and experiences. The breadth and nature of the field of urban agriculture and related educational programs is nuanced, eclectic, highly diverse, and transcends a wide range of contexts. As such, literature was reviewed and summarized into a framework of factors, including scope; approaches and methods; diversity of production and services; systems-driven and holistic; purpose-driven on food and social justice; a community-based food system for the public good. Although numerous authors have discussed the diversity of urban agriculture, these factors were proposed as a framework to continue the dialogue and demonstrate how various authors in this book discussed urban agriculture in their chapters. This book is a collection of authors' wisdom of programmatic and evidence-based experiences in urban agriculture education. Nine chapters are briefly introduced, and their unique contributions are highlighted in the introduction. The chapters provide implications on how educators can inform programmatic engagement in urban agriculture.

Keywords Urban agriculture education · Conceptual framework

N. Knobloch (✉)
Department of Agricultural Science Education and Communication, Purdue University, West Lafayette, IN, USA
e-mail: nknobloc@purdue.edu

I. DeCoito et al. (eds.), *Teaching and Learning in Urban Agricultural Community Contexts*, Urban Agriculture,
https://doi.org/10.1007/978-3-030-72888-5_1

1.1 Overview of Urban Agriculture

Urban agriculture is a growing phenomenon because more citizens are interested in where their food comes from, how it was produced, how it fits the value chain of the food production system, and how it aligns with their personal, social, political, economic, and environmental values. When the United States was a new nation, about 230 years ago, 90% of the U.S. population were farmers and grew their own food (Spielmaker 2018). Today, less than 1% of the U.S. population are farmers. When the U.S. was being settled by immigrants from Europe, urban agriculture was common because people grew their own food, including those who lived in cities (Brinkley and Kingsley 2018).

An example of historical urban agriculture was the infamous fire in Chicago in 1870 when a milk cow kicked over a candlelit lamp. Further, victory gardens provided one-third of the vegetables in the U.S. during WWII (Thornton 2018). Urban agriculture is not a new idea among the scientific and food production interests, but it re-emerged as a more recent phenomenon in the U.S. In the 1970s and 1980s, urban agriculture was recognized as an urban policy issue and strategy for developing urban communities (Thornton 2018).

Initially, urban agriculture declined in the U.S. when technology and the industrial revolution shifted the food production system to rural areas. Through westward expansion of the frontier and the industrial expansion of cities from 1790 to 1890, the food production system concentrated in rural areas (Thornton 2018), which were adjacent to cities and urban markets (Dziedzic and Zott 2012). This separation between producers and consumers occurred overtime and producers sold their commodities to wholesalers and distributors. This separation led to fewer and less frequent interactions with key stakeholders in the food system, which also resulted in consumers not receiving direct information and building relationships with producers.

This separation has resulted in a growing concern of consumers not being agriculturally literate (Chapman and Lindner 2018). Most of the U.S. population is not agriculturally literate (Mercier 2015) and most adults have a limited understanding of agriculture (Stofer and Newberrry III 2017). Most scholars agree there is a need to reconnect producers and consumers because the social relations were eroded or stripped away from the conventional industrial food system in the last 60 years (McClintock 2014). A common thread is that more consumers want to know where their food comes from and how it was produced (Dimitri et al. 2016; Thornton 2018).

The direct connection between producers and consumers is complicated and elusive because of the numerous phases food goes through to reach the consumer (Oosterveer and Sonnenfeld 2012). This widening gap between producers and consumers has mixed responses among consumers. These responses are varied and can range from apathy and not interested (Cotton et al. 2009), embedded beliefs about food choices, routinized food consumption practices (Oosterveer and Sonnenfeld 2012), and limited knowledge of the food system (Stofer and Newberrry III 2017) to being interested, wanting first-hand experiences (Perez and Allen 2007), be engaged in hobby agriculture (Low et al. 2015), creating communities, and activism (Alkon and Norgaard 2009).

Urban agriculture can be a threat to those in the existing industrial food production. Alternative urban food systems are a response to an antagonistic relationship between socially just spaces and neoliberal spaces (Thornton 2018). As the U.S. population becomes more diverse, the interests, expectations, and ways of engaging with agriculture and the food system also become more diverse. The growth of urban populations and the food movement has spawned an interest in urban agriculture. Cooking, teaching, environmental, hobby or aesthetics were the five main motivators for consumers to raise their own food. Although consumers may raise food to save money or have food to survive, the primary motivator was to increase access to diverse, nutritious and culturally appropriate foods (Kortright and Wakefield 2011).

According to the National Gardening Association (2014), consumers' renewed interest in raising food and gardening is based on saving money on food expenses and better tasting and quality food. This suggests that consumers are interested in growing their own food when they do not feel the commercial food system is meeting their needs (Wallinga 2009). In 2020, COVID19 threatened the food supply and there was an increased interest in gardening.

Although some urban agriculture focuses on meeting urban market demands through small-scale intensive production techniques, urban agriculture can focus on social aspects such as food security, community development, and education (Oberholtzer and Dimitri 2016). Therefore, *urban agriculture* is broadly defined as connecting people to the land and their food and bridging cultural traditions and knowledge networks between rural and urban populations (Brinkley 2013; McClintock et al. 2014). More specifically, Mougeot (2000) defined urban agriculture as "an industry located within (intra-urban) or on the fringe (peri-urban) of a town, a city, or a metropolis, which grows or raises, processes and distributes a diversity of food and non-food products, (re-)using largely human and material resources, products and services found in and around that urban area, and in turn supplying human and material resources, products and services largely to that urban area" (p. 11).

There are different types of urban agriculture, which may be residential, allotment, guerilla, collective, institutional, non-profit, or for-profit (McClintock 2014). Urban agriculture is highly diverse in type, purpose and productivity. Urban agriculture can overlap with other descriptors such as the backyard movement (Brinkley and Kingsley 2018), alternative food movement (Thornton 2018), community-based urban agriculture (Nemoto and Biazoti 2017; Ramalingam et al. 2019), reclaiming urban spaces (e.g., pop-up gardens), integrating allotment gardens in buildings (e.g., roof-top gardens; Caputo et al. 2016), intergenerational gardening (Hake 2014), and urban food entrepreneurship (Jones 2018).

Urban agriculture is nuanced, eclectic, highly diverse, and transcends a wide range of contexts, purposes, and values (Caputo et al. 2016; Baycan-Levent and Nijkamp 2005). Urban agriculture can be as simplistic as a patio tomato on the deck of an apartment building in a city or as complex as a social movement regarding food justice of an urban agroecosystem. Urban agriculture is highly diverse and offer a framework to help advance the discussion of thinking conceptually about the definition of urban agriculture. Urban agriculture has various perspectives (Lorenz 2015), meanings, and definitions. Urban agriculture can be nuanced and highly contextualized (Poulsen 2017), depending on a number of factors, including geographic location, scope of operations, approach and purpose. Although numerous authors have discussed the diversity of urban agriculture, the following factors are proposed as a framework to continue the dialogue and demonstrate how various authors in this book discussed urban agriculture in their chapters.

1.2 Framework of Urban Agriculture Education

Location Urban agriculture is "the growing, processing, and distribution of food and other products through intensive plant cultivation and animal husbandry in and around cities" (Bailkey and Nasr 2000, p. 6). Although location has been used to define urban agriculture (Ellis and Sumberg 1998), it is not inclusive of the broader dimensions of urban agriculture, such as the urban economic and ecological system (Mougeot 2000). Urban agriculture is a subsector of the broader food system context and can represented different functions that are embedded in local communities and focus on a different type of food production (Nemoto and Biazoti 2017).

Scope Because of space limitations, urban agriculture in the U.S. tends to be smaller-scale compared to commercialized large-scale agricultural production in rural areas. Accessing land in urban areas can be challenging because land may not be available, have zoning restrictions, be cost-prohibitive, require a large capital infrastructure investment, or be difficult to transfer to growers (USDA 2016). The financial challenges of urban agriculture may also be a function of size and location. According to the USDA (2016, p. 13), "due to limited land availability and higher costs that urban growers often face, it can be difficult for urban growers to make a profit."

Approaches and Methods The approaches and methods used in urban agriculture tend to be more labor-intensive, use vertical space, used integrated approaches such as aquaponics, and involve stakeholders to be part of the production, harvest, and agro-tourism. Urban agriculture is more than producing a commodity or providing fresh food—it can foster experiences such as public engagement, inclusive spaces, democratic participation, intergenerational learning, and social learning about producing and consuming food (Baker 2004; Hake 2014; Levkoe 2006). Urban agriculture tends to highly contextualized and situates itself within the ecosystem and a broader context such as the food system (Poulsen 2017). As such, sustainable production practices play an important role in urban agriculture (Oosterveer and Sonnenfeld 2012; Thornton 2018).

Diversity of Production and Services The benefits of urban agriculture include fresh and local food, public health, physical exercise, green spaces, green jobs, environmental sustainability, rainwater retention, climate change mitigation, neighborhood surveillance, education, and community building and beautification (McClintock 2014). The interest in raising food within urban settings is driven by concerns of food quality, a desire for a more natural lifestyle, building community around food production (Brinkley and Kingsley 2018), and economic development in post-industrial cities (Jones 2018). As industrial production agriculture focuses on becoming more efficient and cost-effective by producing one or two commodities, urban agriculture tends to produce more variety of fresh foods, address environmental injustices, or experiences that integrate with the production system.

Systems-Driven and Holistic Urban agriculture is aligned with values of sustainability, environment, and human health. Urban agriculture tends to look more holistically at the system, in which the food system is situated (Thornton 2018). The interdependence of the food system with the environment and human health provides a broader perspective and decision-making that is not reductionist and technical in nature (Poulsen 2017). Urban agriculture can be a local response to addressing economic, social, and environmental impacts of the conventional and industrialized food system (Thornton 2018). Consumers' concerns about the food system center around five categories: (1) naturalness; (2) food safety; (3) animal welfare; (4) environmental; and, (5) social impacts (Oosterveer and Sonnenfeld 2012). More specifically, Baycan-Levent and Nijkamp (2005) identified five categories of values of urban agriculture and green spaces: (1) ecological values—intrinsic, genetic diversity, and life-support; (2) economic values—market; (3) social values—recreational, aesthetic, cultural symbolization, historical, character-building, therapeutic, social interaction, substitution; (4) planning values—instrumental/structural, synergetic and competitive; and, (5) multidimensional values—scientific and policy.

Purpose-Focused on Food and Social Justice Urban agriculture can be a protective counter-movement in response to large-scale industrial agriculture (McClintock 2014), which can be driven by civic responsibility (McClintock 2014), social equity, environmental justice, and food justice (Thornton 2018). Further, alternative urban food systems address an antagonistic relationship between socially just spaces and existing neoliberal spaces. Urban agriculture food justice advocates view food as a public good rather than a commodity and focuses on equitable distribution rather than profit (Poulsen 2017). The premise of food justice is "people with privilege and wealth should subsidize the less fortunate" (McClintock 2014, p. 148) and making healthy and nutritious food affordable and accessible to unserved, underserved, under-resourced, and underrepresented consumers. Two-thirds of urban farmers in the U.S. had social goals (Dimitri et al. 2016).

A Community-Based Food System for the Public Good Urban agriculture can strengthen communities by engaging the public to participate in social exchanges around issues and projects that require agricultural, food and culinary knowledge and skills. As food is a necessary resource for human survival, health and enjoyment, food can provide concrete examples of the multi-ethnic, social, and economic disparities. Urban agriculture can foster multi-ethnic and multi-generational exchanges to advance the understanding and transfer of human capital of agriculture, food and culinary arts (McClintock 2014). In doing so, urban agriculture can foster civic engagement for inclusive public participation and social learning about the food system (Baker 2004; Levkoe 2006). Urban agriculture can provide transformative outcomes beyond producing food (Travaline and Hunold 2010)—it can develop food citizens for a food democracy (Welsh and MacRae 1998; Hassanein 2003) with an understanding of urban ecological citizenship (Travaline and Hunold 2010) through community engagement, participatory decision-making, environmental stewardship, and local ownership of food system and urban environment. Engaging local citizens in urban agriculture can promote life-long learning, which can occur in formal, nonformal, and informal educational settings (Ochoa et al. 2019). Developing local networks where people can share their experiences can create citizenship engagement.

1.3 Overview of Chapters

This book focuses on evidence-based educational practices of advancing urban agriculture through community-based educational approaches. Whereas, the companion book focuses on educational research studies in the context of urban agriculture. With a focus on practice, this book is a collection of chapters that highlight different educational approaches to engage students to learn about urban agriculture and food systems in a variety of contexts throughout the U.S. and Peru.

All chapters focus on educational practices to engage youth and adults in urban agriculture and the food system. The cross-cutting themes across these chapters touch upon experiential education and learner-centered teaching, community-based engagement and placed-based learning, and multidisciplinary and interdisciplinary curricular approaches. Finally, the resonating theme throughout these chapters are food, community, environment, and people. People are connected to urban agriculture because of food. Why? Food is universal because everyone eats—food is relevant. Food is universal because it connects people across generations, across communities, and across differences—food is participatory. Food is universal because it highlights social injustices, issues of food and environmental justice, and the tensions and conflicts of producing food sustainably for our environment—food is significant.

Deweyian Urban Agriculture Mihye Won and Bertram Bruce described the "Community as curriculum: Dewey's theory of inquiry in the context of an urban agriculture project." In this chapter, the authors shared how citizen science was community science for students enrolled in the Pedro Albizu Campos High School (PACHS) in an ethnically diverse neighborhood in Chicago, Illinois. Students engage in urban agriculture projects by raising plants using a hydroponics system. Through experiential learning, students were motivated to explore ways they could produce food and address local food insecurity in their local community. Based on John Dewey's philosophy of experiential learning, the authors shared examples of how the urban agriculture projects engaged students through community-based inquiry to enhance relevance (relation of activities to learners' experiences), participation (mode of learners' engagement), and significance (impact on learners' individual and communal life). Urban agriculture connects learning to the community and integrates community into the curriculum.

A Framework for Urban Agriculture Helena Farrell described the intersectionality of urban agriculture and interdisciplinary education in her chapter, "Forging research pathways to sustainable farms and food systems with an interdisciplinary evaluative framework for urban agriculture." A comprehensive evaluative framework was developed for an interdisciplinary undergraduate course that was taught at the University of Massachusetts, Amherst. The course focused on teaching students contemporary farming methods and the food system with critical social, economic and environmental considerations. The comprehensive evaluative framework situates the farm system in the context of system dynamics, system design, and outcome and benefits. The evaluative framework guides students to conduct research studies and practice systems thinking of the web of interactions and influences on the food system in which the farming system is embedded. By engaging in this experiential learning designed course and using the evaluative framework, undergraduate students become more aware of careers in the food systems and develop critical, systems, and interdisciplinary thinking skills.

Youth Development Through Urban Agriculture Michelle Wooten and Josh Corlew explained how a nonprofit organization engaged Nashville youth in farming, food choice, and food access issues. In their chapter, "Engaging Nashville's youth in farming, food choice, and food access issues: Two programs by a Nashville nonprofit," the authors described two programs that engaged youth in out-of-school urban agriculture experiences. A non-profit organization hires high school students to serve as teen facilitators for an urban agriculture program known as Crop City and interns for the Communities and Food Internship. Both programs engaged urban youth to learn about the food system in the context of urban agriculture. The authors explained how each program was evaluated and share results of the programs. The programs engaged youth to explore agricultural skills and social justice issues and helped them to better understand how they could address food security in their own communities.

Developing Environmental Identity Through Urban Reforestation Daniela Benavides, Adriana Gonzalez-Pestana, and Joaquín Leguía contributed a chapter on an urban reforestation titled, "Resurfacing environmental identity in coastal Peru." The authors used a Forest for Ancon as an educational tool to engage youth to develop environmental literacy and self-identities. This environmental education project motivated teachers to engage students and teach their lessons in outdoor settings. Specifically, a diverse group of stakeholders planted 500 trees to create a forest located in the city of Lima. The forest was developed using human-centered design principles to provide place-based learning experiences for teachers and students to learn about the importance of trees in the Peruvian coastal desert ecosystem. The authors shared benefits of the project, additional ways the project can be used for new learning possibilities, funding and scale-up options, and how barriers can be addressed for teachers who want to engage in out-of-school learning experiences.

Informal Science Education Program Zev Friedman and Phyllis Katz contributed a chapter on "Permaculture in action: Urban farming as continual science learning." In this chapter, Friedman and Katz shared a story of how an intensive experiential learning program known as Continual Science Learning (CSL) used permaculture (i.e., principles, framework, and system) to engage participants and educators to learn about food production, energy systems, the built environment, and socio-economic perspectives of a permaculture system. The authors discussed how the Permaculture In Action (PIA) program was structured, the curriculum cycle was implemented using six strands of informal learning environments, and the program design was informed by permaculture principles to support transformative learning. This chapter uniquely illustrates how urban agriculture can be operationalized using permaculture as a means to engage participants in holistic informal STEM education to develop and use systems thinking in promoting healthy human ecosystems on planet Earth.

Community-Based PK-12 Food System Education Christopher Murakami and Heather Gillich wrote a chapter on "Learning to become 'good food' educators: Practices and program development of an urban agriculture education organization." In this chapter, the authors shared how the Columbia Center for Urban Agriculture (CCUA) started as a grassroots movement to bring local environmental and food activists together to address social and environmental injustices through food education. The authors illustrated how a healthy soil ecosystem model was used as an educational program framework and shared how local community volunteers and educators engaged PK-12 youth in learning about healthy food through school enrichment presentations, afterschool programs, community gardens, and place-based on-farm learning experiences. Murakami and Gillich integrated evidence-based findings by integrating stories from the founders of the CCUA and provided philosophical and programmatic insights on how to develop community-based urban agricultural education programs for diverse community citizens and youth audiences.

The People's Garden Craig Wilson, Carolyn Schroeder, and Tim Scott described how "The USDA future scientists urban agriculture program" developed youth through real-world research-based experiences in a People's Garden located in College Station, Texas. This chapter described how the People's Garden was created using a pond, Texas pocket prairie, pollinator garden, Monarch Butterfly Waystation, and vegetable garden. The location was reclaimed land on the site of a USDA research facility. The People's Garden created an opportunity for place-based outreach education where students, teachers, parents, and community members could engage in experiential learning at a federal agricultural research facility. The interdisciplinary nature of this program helped students gain research and observation skills, combine language arts with STEM and agriculture, and develop connections with nature through experiential place-based learning.

Systems Approach to Civics Urban Agriculture Kevin Slick and Mila Tewell wrote a chapter on "Forging the farm-to-school connection: Articulating the vision behind food-based environmental education at the Dalton school." In this chapter, they described how they transformed a K-12 private school to teach students more holistic approaches to environmental sustainability using a flexible and adaptive approach to food-based environmental education. Impressively, high school students have a Food Club, which manages a student-run Community-Supported Agriculture (CSA). The CSA serves 100 families and addresses issues of food security in the New York metropolitan area by engaging students and community citizens through political action and community service. The food-based environmental education program develops students' awareness of sustainable stewardship, personal responsibility, and critical and system thinking. The authors provided an overview of best practices and ways to engage high school students to develop a moral responsibility for the health of the land.

Integrated Learning Through Urban Beekeeping Thomas Schmitt, Kristian Demary, and Noah Wilson-Rich shared a unique and innovative perspective on using "Urban beekeeping as a tool for STEAM education." In this chapter, the authors shared how urban beekeeping can be used as a way to engage K-12 students through project-based learning, and coined the term, "apiary pedagogy." The authors provided an overview of the practical uses and applications of bees, and technical knowledge regarding bee nests, hive structure, pathogens, parasites, and phenotypic advantages. Pedagogically, this chapter demonstrates how bees can be a means to engage students in STEAM learning. Urban beekeeping in the context of urban agriculture can help connect students to nature and the environment. Engaging in urban beekeeping project-based learning can help students develop critical thinking skills, decision-making, creative design, and career awareness.

1.4 Summary

Urban agriculture as broadly defined using a systems approach across various domains in local contexts provides unique opportunities to engage youth and adults in learning together about the food system. This book highlights examples of how educational and community organizations provide relevant, engaging, and holistic learning experiences that connect youth, educators, and stakeholders to food, agriculture and the environment. Hopefully, you find these chapters insightful in illustrating how experiential learning, context-based experiences, community-based approaches, systems thinking, and integrated learning can engage youth and adults to be engaged citizens in the urban food system.

References

Alkon, A. H., & Norgaard, K. M. (2009). Breaking the food chains: An investigation of food justice activism. *Sociological Inquiry, 79*(3), 289–305.

Bailkey, M., & Nasr, J. (2000). De Brownfields a Greenfields: Producir el alimento en ciudades norteamericanas. *Noticias de la seguridad del alimento de la comunidad*

Baker, L. E. (2004). Tending cultural landscapes and food citizenship in Toronto's community gardens. *Geographical Review, 94*(3), 305–325.

Baycan-Levent, T., & Nijkamp, P. (2005). Evaluation of urban green spaces. In D. Miller & D. Patassini (Eds.), *Beyond benefit cost analysis: Accounting for non-market values in planning evaluation* (pp. 63–87). Aldershot: Ashgate.

Brinkley, C. (2013). Avenues into food planning: A review of scholarly food system research. *International Planning Studies, 18*(2), 243–266.

Brinkley, C., & Kingsley, J. S. (2018). Urban agriculture. In *Advances in Agricultural Animal Welfare* (pp. 241–257). Woodhead Publishing. doi: https://doi.org/10.1016/B978-0-08-101215-4.00013-4

Caputo, S., Schwab, E., Tsiambaos, K., Benson, M., Bonnavaud, H., Demircan, N., & Pourias, J. (2016). *Emergent approaches to urban gardening* (pp. 229–253). London/New York: Urban allotment gardens in Europe. Routledge.

Chapman, D. L., & Lindner, J. R. (2018). *Teacher perception of the Georgia middle school agricultural education curriculum*. Proceedings of the Southern Region Conference, American Association of Agricultural Education (pp. 95–109). Jacksonville, Florida

Cotton, C. P., Hashem, F. M., Marsh, L. E., & Dadson, R. B. (2009). Broadening perspectives: Educating underrepresented youth about food agricultural sciences through experiential learning. *NACTA J, 53*(4), 23–29.

Dimitri, C., Oberholtzer, L., & Pressman, A. (2016). Urban agriculture: Connecting producers with consumers. *British Food Journal, 118*(3), 603–617.

Dziedzic, N., & Zott, L. (2012). *Introduction to urban agriculture: Opposing viewpoints. Urban Agriculture*. Detroit: Greenhaven Press.

Ellis, F., & Sumberg, J. (1998). Food production, urban areas and policy responses. *World Development, 26*(2), 213–225.

Hake, B. J. (2014). *What grows in gardens? In learning across generations in Europe* (pp. 155–166). Rotterdam: SensePublishers.

Hassanein, N. (2003). Practicing food democracy: A pragmatic politics of transformation. *Journal of Rural Studies, 19*(1), 77–86.

Jones, J. C. (2018). *Urban food entrepreneurship, governance, and economic development in the post-industrial cities of Newark, New Jersey and Dayton, Ohio*.

Kortright, R., & Wakefield, S. (2011). Edible backyards: A qualitative study of household food growing and its contributions to food security. *Agriculture and Human Values, 28*(1), 39–53.

Levkoe, C. Z. (2006). Learning democracy through food justice movements. *Agriculture and Human Values, 23*(1), 89–98.

Lorenz, K. (2015). Organic urban agriculture. *Soil Science, 180*(4/5), 146–153.

Low, S. A., Adalja, A., Beaulieu, E., Key, N., Martinez, S., Melton, A., Perez, A., Ralston, K., Stewart, H., Suttles, S., Vogel, S., & Jablonski, B. B. (2015). Trends in US local and regional food systems: A report to Congress. Retrieved from: https://www.ers.usda.gov/webdocs/publications/42805/51173_ap068.pdf

McClintock, N. (2014). Radical, reformist, and garden-variety neoliberal: Coming to terms with urban agriculture's contradictions. *Local Environment, 19*(2), 147–171.

McClintock, N., Pallana, E., & Wooten, H. (2014). Urban livestock ownership, management, and regulation in the United States: An exploratory survey and research agenda. *Land Use Policy, 38*, 426–440.

Mercier, S. (2015). *Food and agriculture education in the United States*. Washington, DC: AGree.

Mougeot, L. J. (2000). Urban agriculture: Definition, presence, potentials and risks, and policy challenges. *Cities Feeding People Series: Report 31*. International Development Research Center (IDRC). Available: http://idl-bnc.idrc.ca/dspace/bitstream/10625/26429/12/117785.pdf

National Gardening Association. (2014). *Garden to table: A 5 year look at food gardening in America*. Williston: National Gardening Association.

Nemoto, E. H., & Biazoti, A. R. (2017). Urban agriculture: How bottom-up initiatives are impacting space and policies in São Paulo. *Future of Food: Journal on Food, Agriculture and Society, 5*(3), 21–34.

Oberholtzer, L., & Dimitri, C. (2016). *Urban agriculture in the United States: Baseline findings of a Nationwide survey*. Butte: National Center for Appropriate Technology.

Ochoa, J., Sanyé-Mengual, E., Specht, K., Fernández, J. A., Bañón, S., Orsini, F., Magrefi, F., Bazzocchi, G., Halder, S., Martens, D., Kappel, N., & Gianquinto, G. (2019). Sustainable community gardens require social engagement and training: A users' needs analysis in Europe. *Sustainability, 11*(14), 3978.

Oosterveer, P., & Sonnenfeld, D. A. (2012). *Food, globalization and sustainability*. London: Routledge.

Perez, J., & Allen, P. (2007). *Farming the college market: Results of a consumer study at UC Santa Cruz*

Poulsen, M. N. (2017). Cultivating citizenship, equity, and social inclusion? Putting civic agriculture into practice through urban farming. *Agriculture and Human Values, 34*(1), 135–148.

Ramalingam, L., Sharifuddin, J., Mohamed, Z. A., & Ali, F. (2019). Motivation and satisfaction of volunteers for community-based urban agriculture programmes. *International Social Science Journal, 69*(231), 49–62.

Spielmaker, D. M. (2018). Growing a *Nation Historical Timeline*. Retrieved from https://www.agclassroom.org/gan/timeline/index.htm

Stofer, K. A., & Newberrry, M. G., III. (2017). When defining agriculture and science, explicit is not a bad word. *Journal of Agricultural Education, 58*(1), 131–150. https://doi.org/10.5032/jae.2017.01131.

Thornton, A. (2018). *Space and food in the City : Cultivating social justice and urban governance through urban agriculture*. Cham: Palgrave Pivot. https://doi.org/10.1007/978-3-319-89324-2_1.

Travaline, K., & Hunold, C. (2010). Urban agriculture and ecological citizenship in Philadelphia. *Local Environment, 15*(6), 581–590.

United States Department of Agriculture. (2016). *Urban agriculture toolkit*. Washington, DC: United States Department of Agriculture.

Wallinga, D. (2009). Today's food system: How healthy is it? *Journal of Hunger & Environmental Nutrition, 4*(3–4), 251–281. https://doi.org/10.1080/19320240903336977.

Welsh, J., & MacRae, R. (1998). Food citizenship and community food security: Lessons from Toronto, Canada. *Canadian Journal of Development Studies/Revue canadienne d'études du développement, 19*(4), 237–255.

Chapter 2
Community as Curriculum: Dewey's Theory of Inquiry in the Context of an Urban Agriculture Project

Mihye Won and Bertram C. Bruce

Abstract Urban agriculture programs are recognized as an effective way to bring students' cultural funds of knowledge into their school-based science learning, and in turn, to use the school-based learning to effect changes in the community. However, despite their potential to engage students in meaningful learning and to break the boundaries between school and community, many such programs are poorly integrated into the school-based science curriculum. In this study, we describe an urban agriculture project that was systematically integrated into high school science teaching and supported by the whole school community, later contributing to community action in the neighborhood. Employing John Dewey's theory of inquiry as the analytical framework, we discuss the educational implications of the urban agriculture project and examine the goals of education and the value of the urban agriculture program, in terms of the growth of students and community.

Keywords Community-based education · Inquiry-based teaching and learning · Urban agriculture · John Dewey · Theory of inquiry · School curriculum · Community action

M. Won (✉)
School of Education, Curtin University, Perth, Australia
e-mail: mihye.won@curtin.edu.au

B. C. Bruce
School of Information Sciences, University of Illinois, Champaign, IL, USA
e-mail: chip@illinois.edu

I. DeCoito et al. (eds.), *Teaching and Learning in Urban Agricultural Community Contexts*, Urban Agriculture,
https://doi.org/10.1007/978-3-030-72888-5_2

2.1 Urban Agriculture as a Link to Connect School and Curriculum

Urban agriculture programs are recognized as an effective way to bring students' cultural funds of knowledge into their school-based science learning (Fusco 2001). Exemplifying place-based education, they provide opportunities for students to analyze the economic and political decision making of their locales (Gruenewald 2003). Most urban community garden organizers value the education of younger members as an essential component (Armstrong 2000; Weissman 2015). They encourage youth to participate in building healthy and democratic communities as citizen scientists (Mueller and Tippins 2012). However, despite their potential to engage students in meaningful learning and to break the boundaries between school and community, urban agriculture programs are rarely integrated fully into school-based science curriculum. Instead, they remain a short-term, one-time event that science teachers may or may not incorporate into their lessons.

In this study, we describe an urban agriculture project that was systematically integrated into school operations and supported by the whole school community, later becoming part of community action in the neighborhood. Students were citizen scientists in the sense that they collected and analyzed data and learned about natural phenomena along the way. Going beyond many citizen science projects, they also generated the questions for inquiry, interpreted their findings, and applied them to address community needs. In these ways, citizen science truly became community science.

We describe how the project originated, what was actually implemented, and its implications. Our main goal is not to evaluate the effectiveness of this particular project. Rather, we intend to examine the value and possibilities of urban agriculture programs in terms of the educational growth of students and community, and to illustrate how John Dewey's theory of inquiry can be used as a framework to analyze community-based urban agriculture education programs (Enfield 2001; Won 2009).

2.2 John Dewey's *Theory of Inquiry* as Analytical Framework

We adopted John Dewey's theory of inquiry to capture important aspects of the community-based school science curriculum and activities. Dewey's ideas have been influential in science education over the years, often with an emphasis on hands-on learning or experimentalism. Yet, other aspects of Dewey's ideas are often overlooked, in particular, his ideas about the fluidity of inquiry problems, the situated nature of inquiry, reflection on hindrances to inquiry, and inquiry as a participatory practical action.

In his *Logic: Theory of Inquiry* (1938b), Dewey describes how our understanding of the world develops through a process of inquiry. Knowing is a process of resolving indeterminacy in a situation, rather than one of accumulating facts. From this perspective, learning means developing and reconstructing our experiences to act upon significant aspects of our lived experience in a productive way. An important implication for curricula is that schools and other educational institutions need to provide constructive learning environments, where students bring in their significant experiences to explore, experiment, and transform so as to instigate progress on both individual and social challenges (Dewey 1900/1976, 1916/1980).

We focus on three key aspects of Dewey's theory: *relevance* (relation of activities to learners' experiences), *participation* (mode of learners' engagement), and *significance* (impact on learners' individual and communal life).

To start the process of learning, learners need to realize the *relevance* of the problem or activity (Dewey 1938a, b). This component corresponds to the intention or intended goals of educational activities. Dewey (1938b) wrote, "To set up a problem that does not grow out of an actual situation is to start on a course of dead work" (p. 112). When the activity relates to learners' lived experiences in a meaningful way, they put genuine effort to figure out and resolve the problematic situation. Because students' lived experiences determine the quality of a school activity for them (Dewey 1902/1976, 1938b), teachers need to endeavor to provide more educative experiences based on their deep understanding of students' experiences.

In the process of learning, the mode of engagement should be collaborative and reflective *participation* (Dewey 1916/1980, 1938a). The alignment of the means and the ends is central in Dewey's view on education. As students are engaged in a problem-solving activity of personal and social importance, the process needs to reflect the educational aims we want to instill in them. This mode of collaborative and reflective participation applies not only to students' interactions amongst themselves, but also in relation to the teacher(s), and even to community members outside of the school.

The outcome of the learning activities can be referred to as the *significance* of the learning. Because Dewey conceptualized learning as a dynamic interaction between the learner and the situation at hand, the outcome of learning is not limited to the knowledge the learner attained. Rather it is understood as a transformation of the learner and the problematic situation. "Experience does not go on simply inside a person .... Every genuine experience has an active side which changes in some degree the objective conditions under which experiences are had" (Dewey 1938a, p. 22). By transforming in a meaningful way, the problematic situation and our interaction within it, learners experience significant learning.

For Dewey, then, inquiry begins in experience (relevance), requires active engagement (participation), and then returns its results to experience (significance). Unfortunately, formalized learning often reduces this three-stage process to the middle part of problem solving, and furthermore reduces that to limited or passive participation. The transformative power of inquiry is thus reduced, and it is less likely to have enduring value for the students.

Table 2.1 Learning component checklist

Key Questions
Relevance: Connection to students' experiences
How was the inquiry activity initiated and understood in relation to the students' own personal and social experiences? How did the activity utilize the cultural, experiential resources of students?
Participation: Inquiry with reflection and collaboration
How did the students contribute to the design and operation of the activity? How did the students collaborate with others? How reflectively did the students examine the progress of the inquiry?
Significance: Meaningful transformation of the situation
What changes did the inquiry effect on the problematic situation? How did the activity help students participate in their life matters better? What further inquiry did the activity initiate?

In order to make these components of learning more visible in our analysis as well as to provide practical guidance for educators, we chose three questions for each (Table 2.1). The list of questions is used to capture the essential features of learning activities. Based on this list, we analyzed the urban agriculture project at an alternative high school in Chicago.

2.3 Urban Agriculture Project at PACHS

2.3.1 School and Community Context

Pedro Albizu Campos High School (PACHS) is located in an ethnically diverse neighborhood in Chicago, with a large Puerto Rican population. Part of the Alternative Schools Network, the school has a reputation as a different kind of school, where students' concerns are actively brought in to the school curriculum, and students are supported to develop their cultural identities through community-based activities (Antrop-Gonzalez 2003; Berry and Cavallaro 2014; Bruce 2008; Johnson 2005).

The strong commitment towards its community is systematically built into the school's operations (Flores-González et al. 2006; Ramos-Zayas 1998). That commitment is apparent in the location of the school—the Puerto Rican Cultural Center, where people come to plan and organize community activities concerning health, legal, psychological, and financial issues. The majority of teachers at PACHS are active community organizers, living within a few blocks of the school (Antrop-Gonzalez 2003). They collaborate with other community members to plan and execute a variety of school activities and community actions, even during summer holidays (Bruce 2008; Bruce and Bloch 2013). They commit themselves to various community activities outside school, such as participating in protests, planning cultural events, or cleaning up the community (Antrop-Gonzalez 2003). There is a regular town hall-type meeting so students can talk about the learning process within the school and how it might be improved. Due to the school's effort to provide cultural affirmation and a caring relation to students' lives, the dropout rate has dramatically decreased. Now, most students successfully graduate from high school and many of them go on to college (Bruce 2008). The school states its approach on its website (https://pachs-chicago.org):

> We offer basic skills training with activities in technology, urban agriculture, media production and communications, critical art, and participatory action research, as well as college preparatory coursework through an ongoing partnership with Wright College. Classroom instruction is primarily project and problem based with an emphasis on developing higher order thinking skills of inquiry and analysis in place of rote memorization. Planning for the years to come, our curriculum will have social ecology as a conceptual centerpiece, framing student learning around three experimental stations: a hydroponics-based greenhouse on the roof of PACHS; soil-based community gardens in and around Humboldt Park; and tropical soil-based agriculture on the island of Puerto Rico.

2.3.2 Development of the Urban Agriculture Project at PACHS

The school's community-based curriculum is enacted in part through urban agriculture projects, located within community action movements (Krasny and Tidball 2009). Urban community gardeners in the US are often recent migrants from developing countries, with extensive practice-based knowledge in farming (Krasny and Tidball 2009). Considering those two aspects, it is no surprise to see an urban agriculture project at PACHS as a way to enlist community resources and to address community concerns.

The urban agriculture project grew organically, with students' realization of the problem accompanied by the teacher's careful planning and support. When the new school year started one September, students saw thriving plants in a hydroponic garden with two-liter soda bottles on the windowsill of their science classroom. The students wanted to start a hydroponic garden project of their own. The teacher and students discussed ways to build a larger hydroponic unit, purchased an aquarium, pipes, and a pump, and made a hydroponic garden unit. As they tended the plants in the hydroponic unit, they learned about hydroponic gardening in comparison to soil-based gardening—how plants grow without soil, the benefits and drawbacks of this method of gardening, etc. They also talked about the possibility of beautifying the community with flowers.

One day, the class discussed different issues in the community, such as gang activity and domestic violence. As they talked about gangs—the nature of the problem, contributing factors, and what could be done—they realized that people joined because gangs provided some of their basic needs—food, housing, and safety. The students reasoned if they could contribute to providing those basic needs to the community, people would not have to be involved in gangs as much. The science teacher then narrowed down the discussion to the food issue in the community. The students questioned whether it was a matter of not having access to affordable, healthy, and authentic food in the community. If so, how could they help address that issue? The students thought that if people could grow their own food, they could generate their own supply and not worry about how and where to get their food. What could they do to help increase the availability of fresh food to the local residents? They wanted to explore this possibility.

Based on their experience with simple hydroponic gardening, students started researching and comparing a variety of ways to supply fresh vegetables to the community. The community organizers and higher education institutions were brought in to provide support in researching and identifying necessary resources for the students. The students spent hours doing research in small groups. Each group established goals, investigated how to implement their plans, and scrutinized what real impact they would have on the community. For example, a group of students proposed to build a rooftop garden on the school building to increase the vegetable supply. In order to decide which system they wanted to put in place, they examined the productivity of soil-based gardening versus hydroponic gardening for the vegetables they wanted to grow. Another group wanted to convert empty lots into community vegetable gardens. They investigated the cost and the environmental impact of urban vegetable gardens. Another group wanted to have a greenhouse at the school for a year-round supply of fresh vegetables, and they examined how to build it and how to secure funds for it. Yet another group wanted to start a farmers' market to stimulate local vegetable growing. They studied the economic impacts of vegetable growing and the farmers' market for the community. They also looked into the possibility of bottling the vegetables, especially making sofrito sauce used as a food base, and investigated the social, financial, and biological aspects of food processing.

For their research/action projects, students used two plots at the community garden to investigate what was involved in growing vegetables in and around the community. To test out their vegetable growing schemes, the students grew several vegetables with important ethnic significance, such as green chilies and cilantro, which are ingredients for salsa sofrito. The teacher and students faced various obstacles along the way, however. Growing Puerto Rican vegetables in Chicago weather without extensive farming resources was challenging, and the anticipated yield did not materialize. As time passed, some students' enthusiasm waned.

While the students did not succeed in supplying fresh ethnic vegetables to the community, the project worked as a platform for the community to talk about health and food security. As students were tending the garden, the community members came out to share their concerns over food security and poor health situations and contributed to weeding and watering of the plants. Students also made a Puerto Rican dish with the vegetables and shared the food at a community club where they presented their research results. Many community members came out to the event and supported the students' projects.

2.3.3 *Researchers' Roles*

As participant observers, our visits to the school involved the Biology class, but also History and after-school programs. We visited related sites, such as the community garden and the farmer's market. We also conducted informal interviews with students, teachers, community leaders, and other university-based collaborators, such as Michelle Torrise (2010), who played a more active implementation role. In addition to relying on direct observation, we crucially drew from a growing body of literature about the community, the high school, and projects there, including the urban agriculture project. This paper synthesizes these findings, seeking to understand its implications for curriculum and instruction.

2.4 Analysis of Urban Agriculture Project Through Dewey's Framework

2.4.1 *Relevance*

2.4.1.1 Initiation of Inquiry Based on Students' Experiences Outside School

The students initiated this project after recognizing fundamental problems in their community, such as gangs, poverty, food security, and poor health conditions. Their understanding of the problems was concrete and real because they had or knew someone who was affected by these problems. The issues, threatening the integrity of their daily lives, were significant and relevant parts of students' experiences, and the students were motivated to do something about these problems to change their living conditions.

In this problem recognition process, the family-like school environment helped students share their significant experiences with teachers and contributed greatly to the launch of the urban agriculture project. The teacher was able to turn the students' initiatives into a series of activities so that they learn how they could use scientific investigation to address those issues. But if the students had not been able to openly discuss gangs and their living conditions with the science teacher, the project could not have started.

Often science teachers focus on 'science matters' such as photosynthesis and controlling variables because they regard their primary responsibility is to pass on scientific knowledge and process skills. They do not talk openly about students' nonacademic issues in class, such as gangs and food security. This does not imply a lack of care. But, the nonacademic issues are regarded as matters for the school counselors or outside science curriculum because of limited class time, privacy concerns, or the uncertainty of success in such discussions.

At PACHS, however, the science teacher was deeply involved in the students' lives, along with the school counselors and other subject teachers at the school. The teachers knew many students were struggling with serious life challenges, as they themselves had experienced as former students of the school or as community members. The teachers thus conceptualized teaching as a means to help a variety of the students' life challenges. They actively encouraged students to talk about important matters in their lives—through school assemblies, individual counseling, and discussions in science class. Because of the teachers' strong engagement and their understanding of students' life experiences, the students regarded their teachers as one of them, people whom they could confide in about important matters.

Understanding students' nonacademic, social experiences was not only a means for the PACHS teachers to build rapport with students; it was the foundation for their teaching, a base from which they could encourage students to investigate significant problems of their own and make science learning meaningful. Understanding students' experiences as a critical component of creating meaningful science learning environments, the teachers at school re-conceptualized the boundary of science teaching to include the enjoyments and challenges in students' lives.

2.4.1.2 Lesson Planning with the Community's Resources

While the students' inquiry appears to have arisen spontaneously, the teacher planned and prepared a series of learning activities in advance to guide the students effectively and link the activities with the science curriculum. Based on the students' interest and excitement in growing plants from hydroponic gardening from the previous school year, the science teacher met with the community members to plan related activities for an urban agriculture and social ecology project, a project they had been thinking about doing before. The community members and the teacher understood that doing an urban agriculture project would provide an opportunity for the students to participate in a community action and to understand their community better, in terms of their culture, healthy diet habits, and environmental issues. They were aware of the international urban agriculture movements to increase food security for low-income urban neighborhoods, make the urban environments greener and better harmonized with nature, and help improve the poor health conditions (obesity, diabetes, and heart disease) of the residents (Corrigan 2011).

Over the summer break, the plan for an urban agriculture project started to take shape. With the help from higher education institutions, a wide range of curricular activities and further investigation areas were identified. To enable a community garden project, a funding proposal was submitted to buy the necessary gardening equipment.

Although the teacher had a tentative plan for the project, he did not hastily present the topic to the students and pressure them to get involved. Rather, he carefully staged the urban agricultural project. He first introduced a rudimentary hydroponic garden unit to bring out students' interest in gardening and urban ecology. As the students enjoyed the observation of plants and wanted to test hydroponic gardening versus soil-based gardening, he initiated a discussion of community issues with students. From the many issues plaguing the community, he narrowed it down to food insecurity and steered the discussion to what they could do about it. He demonstrated the link between food security issue and students' interest in gardening. Then, he helped them inquire an urban agriculture project further, in relation to science, environment, and community development.

As the teacher knew the topic was relevant and prepared a certain amount of educational resources, he was able to bring together confidently the students' individual interests in gardening, the community's need for food security, and the school subject matter. Dewey (1902/1976) knew that the curriculum needs to have an organic connection with the students' experiences, and teachers need to provide a rich environment for students to develop their individual interest in harmony with their social interest. Although the teacher was not explicitly following Dewey, he artfully brought forth students' interests and facilitated the engagement in a community-based inquiry project so that students could connect with both the subject matter and the community.

2.4.2 Participation

2.4.2.1 Students' Participation Modes Through Various Challenges

Many students enthusiastically participated in seeding, planting, watering, and tending the plants in the community garden. When they went out to the garden, they would get their hands dirty and observe the plants closely. When they found something growing, they got excited, and continuously asked questions. They were curious about many things and had many discussions.

Different from the initial small hydroponic gardening, there were many unanticipated challenges in growing the plants in the community garden in Chicago. For example, the students found that bugs and diseases were eating the plants. The plants required more watering than they had expected. There was no irrigation system in place, and a drought occurred, making it necessary for the students to haul buckets of water to the site. The plants they chose required not only water but a lot of sun. Because the weather in Chicago was quite different from Puerto Rico, the plant yield was poor. In addition to the general issues of gardening, the students faced other challenges. Some fruit in the community garden was stolen one day, and some plants were vandalized on another day. The students were upset and discouraged. They were not able to understand why some people would do such things when they were trying to do something good for the community.

Those challenges thwarted the initial inquiry efforts. They also presented valuable learning opportunities, generating a series of small inquiry projects. The students discussed how to diagnose and treat the diseases, the benefits and the dangers of using pesticides, how to manage the soil, and ways to increase yields. They learned about water shortage and changing drought patterns, and how they could help deal with other global environmental issues. They talked about how to productively manage collaboration and how to negotiate with other members of the community to find resolutions.

When solving a real-world problem, we naturally face small and big challenges. The upside of having such challenging experiences at school was that the students had their teacher and friends to discuss and find resolutions more collaboratively and systematically, thus building the knowledge and skills necessary for community-based scientific inquiry.

The majority of the students continued to participate in the project with enthusiasm despite the challenges. However, some students lost interest over time. To re-engage those students, the teacher assigned them a variety of physical tasks for tending the garden, such as hauling water or pulling weeds.

There could be multiple explanations for this reduced participation. The challenges could have been simply too complex and lengthy for the students to keep motivated. The students could have been too pre-occupied with their personal problems to stay focused.

Referring back to Dewey's theory of inquiry, however, leads us to ask what inquiry those non-participants were engaged in. Because learners possess diverse experiences, their interactions with the situation could be different from what the teacher and students themselves initially planned. Some students might have had different interests or immediate personal concerns that were not aligned with the urban agriculture project, but there was not enough allowance for individualized inquiries for those students whose focus or needs were not aligned with the whole class project.

The teachers at PACHS were effective at communicating with students about their communal concerns and deciding on which actions to take with students. It was the intention of the teachers to help students collaborate with one another to reach a common goal. However, the majority of the inquiry projects at school were organized as a series of whole group activities, without allowing individual students to take their own personal approaches.

While the urban agriculture project started out organically, the process of completing the task did not accommodate the different modes of participation of students, partly because the school did not have the resources that would allow individualized inquiries, such as at Dewey's Lab School (Mayhew and Edwards 1936/1965; Tanner 1997). Also, as a teacher, it is often difficult to let go of the control of class activities, especially when students' inquiries are different from our planned activities and we do not fully know where their inquiries would lead them (Osborne 1998). Dewey urged teacher's openness of mind toward the students' ways of addressing a situation: "The teacher who does not permit and encourage diversity of operation in dealing with questions is imposing intellectual blinders upon pupils—restricting their vision to the one path the teacher's mind happens to approve" (Dewey 1916/1980, p. 182).

2.4.2.2 Collaboration Beyond the School

The school personnel and the students always regarded the immediate community as their primary partner, audience, and resource. The urban agriculture project was no exception. Not only did they start out the project based on a community problem, they also surveyed and worked with other community members to determine the needs and means of vegetable gardens. They used a community resource, the community garden, as their field site to experiment with gardening. Members of the community also helped with weeding and watering the plants in the community garden. Students presented their research results in front of the community outside school.

The students realized that this learning was not only for their benefit; it was to become the groundwork for a larger community project. Community members would face the same problems if the students' campaign succeeded and the community started growing their own vegetables. The students' discussions and resolutions of the encountered problems could be used as a valuable knowledge resource for the community to draw upon, as the students were doing the initial explorative work for the community's gardening project. For the students at PACHS, the community was the central and integral part of their curriculum.

The PACHS teacher and students also worked with outside organizations for this project. The collaboration with higher education institutions enriched the activities in multiple ways—they brought in human and material resources to enable the school to plan and execute the urban agriculture project and related activities. For example, one of the university partners helped the teacher search resources for the hydroponic garden unit and examined the potential learning activities in relation to the state science education standards. A graduate assistant also helped with the students' research and amalgamating their findings during the multiple interdisciplinary projects (Torrise 2010). For the curriculum development, a nearby college offered a course based on an advanced food and health study for the PACHS students.

2.4.3 Significance

2.4.3.1 Changes in the Situation

The urban agriculture project started out with the recognition of food inequality and insecurity as a communal problem and aimed to afford the community more control over the production and distribution of healthy, nutritious, and inexpensive food. As research studies have found, community garden projects often have the recognizable health benefits, such as supply of fresh vegetables, and providing physical exercise opportunities (Armstrong 2000).

Although the students and the teacher worked diligently on the project, the immediate effect on the community was not clearly visible. For example, at the beginning of the project, the students thought they could grow and sell 'tons of vegetables' to the immediate effect. Because they were inexperienced in farming and the community garden plots were small, they were not able to achieve that goal. In fact, they sold the vegetables after harvest at a farmers' market one day and earned only $20. It was a humbling experience for the students. Farming was not an easy job, and their enthusiasm alone did not make fundamental changes for the community.

Such difficulties and disappointments might be the very reasons why most teachers do not engage students in tackling critical questions of community, but direct them instead to a simulated problem of the outside world. It might have been prudent to provide a realistic view on the expected outcome of the project to the students at the outset. However, it is doubtful that students would have been emotionally and intellectually involved in the process without such anticipation to solve a real community problem. It is also very important for students to experience 'non-success' and identify ways to improve the outcomes.

While they did not achieve many visible changes, they did achieve what numerous other community garden projects fail to achieve—politicizing the food inequality and understanding social and economic issues around gangs. Weissman (2015) notes the majority of people involved in urban agriculture programs in the U.S. recognize that the youth education component is the most important aspect of their programs. However, he criticizes that many urban agriculture programs focus on the importance of individuals making smart choices for healthy eating habits, rather than collective, critical perspectives on food inequality as the main message of the programs.

The teacher and students at PACHS continue the work at the garden and research additional questions to address social food inequality. The students have investigated the soil—its pH, nutrient compositions, earthworms as fertilizer, making compost, etc.—in order to increase the yield. The teacher continued developing and revising a curriculum for new students, integrating the urban agriculture theme in the context of social ecology. It was an interdisciplinary curricular theme to extend their inquiry. Many class activities and discussions were constructed based on the urban agriculture theme for the students to gain an environmental, economic, and sociological understanding of the community.

2.4.3.2 Students' Experience as the Goal

Science educators consider using everyday phenomena to explain science concepts. However, such a strategy only focuses on showing a small part of the link between science and everyday experiences. The urban agriculture project at PACHS, in contrast, aimed at a different kind of appreciation of science in students' daily experiences. The meaning making of their life experiences was the driving force and the goal of the project. The experiences were closely examined to identify the problem, actively explored to effect change, and critically reflected upon to make further changes. Students' experiences were not just a device to teach science concepts.

Rather, they were the goal of the students' inquiry-based learning. Science was a resource to help them understand a community challenge and act on it.

This community action approach to science pedagogy is not widely employed. However, if we define scientific literacy as a skill set to make sense of the world and live effectively in society, we need to provide opportunities for them to learn and use science in/for their lives (Eisenhart et al. 1996; Roth 2007). As the urban agriculture project demonstrated, a community-based, problem-solving project yields such possibilities with its strong connection to students' everyday experiences and with the alignment of individual and social interests (Burr 2001; Cummings 2000; Giles and Eyler 1994; Kahne and Westheimer 1996; Saltmarsh 1996). Rather than focusing exclusively on transmitting science concepts or skills, science educators need to seriously consider community-based learning opportunities so that students experience the benefits and consequences of science learning on their lives.

2.5 Discussion

2.5.1 Urban Agriculture Project at PACHS: Investigating a Significant Problem for Community Action

The urban agriculture project at PACHS cannot be understood separately from the school's involvement in various community actions because the school had been functioning as the social center of the community (Benson et al. 2007). The integration of the community's needs into school learning was systematically built in the school atmosphere. This urban agriculture project was also drawn from the community's needs and the students' experiences. Because it dealt with a significant problem in students' lives, and most of the students enthusiastically participated in the project, it would display relatively high relevance.

The project, however, did not prove to be straightforward. The students faced a series of difficulties, and they had to devise ways to address them. The problem-solving process was challenging, but it turned out to be a learning opportunity for them to study plants, farming, and the environment more closely. It also drew collaboration with other community members and outside organizations. Although the project involved multiple activities to reflect emerging questions, there were still some nonparticipating students—maybe because they were not able to investigate different questions or contribute to the class in a different way. If there had been a more delicate balance between individual and group projects in the class, they could have engaged in more meaningful learning through productive participation.

While the project was much anticipated, it did not yield many visible changes by the end of the year. This was not because the students and teachers did not work hard on it, but the question had a different quality, and thus the transformation of the situation could not result in a similar form. Although the project did not immediately improve the quality of life for the community members, the students' efforts contributed to the community's knowledge resources as part of an ongoing continuous endeavor to build a better community and improve the life of community members. The students carried on their inquiry with refined questions, and the teachers developed more curricular resources from the project.

The power of relying on students' lived experience is evident in the urban agriculture project at PACHS. Students begin with immediate engagement in learning and build upon their ordinary experience. Moreover, as the project evolves into community action, they can see the consequences of otherwise abstract conceptions.

However, an obvious problem confronts anyone interested in the broad range of students and schools: If the approach relies on specific lived experiences, isn't it doomed to be location specific and non-transferable? One is reminded of Gale's (2006) critique of Dewey's theory of inquiry. He argues that since Dewey's theory relies upon the notion of situation, which he conceives as unique and therefore ineffable, collective inquiry, transferability, or even general knowledge are impossible.

The transferability problem exists even within PACHS. Can the lessons of the urban agriculture project be applied to other topics? Can they even be extended to next year's students? The answer here is twofold. First, each situation is unique and one should not expect to examine what PACHS did with urban agriculture in a given year and replicate it in another situation, especially one that differs in crucial ways, for example, in a rural setting, with younger or older students, or with different cultural histories. On the other hand, the general process of building science curricula out of community needs, knowledge, and values does seem replicable. This has been well-documented in a variety of settings (e.g., Bouillion and Gomez 2001; Moll et al. 1992; Oakes and Rogers 2006).

2.5.2 Adopting Dewey's Theory for Curriculum Innovations

Using Dewey's theory of experience and inquiry, this paper examined the educative value of an urban agriculture project at PACHS, along the way questioning some contemporary goals of school education in terms of the growth of students and community. Stretching the limits of conventional school practices, the students and teachers at PACHS actively embodied Dewey's educational visions, in terms of relevance, participation, and significance. They investigated students' shared concerns in the project and revised the school curriculum to encourage students' integrated understanding of the environmental, economic, and sociological issues of themselves and of the community. They collaborated with one another, with other community members, and with academic faculty members from local higher education institutions. They also attempted to effect significant changes in their lives and in the community, by trying out multiple strategies to make the urban agriculture project successful and initiated discussions with other community members on poverty, food security, and health.

Tempted to applaud the innovative approach at PACHS, readers might still wonder how feasible it is to adopt in mainstream schools, especially in the era of accountability and high stakes assessments. How dependent is it on the school's unique history and practices?

Deweyan inquiry as described in this study can be frustrating (Boostrom 2016; Cunha 2016). It may seem removed from immediate teaching concerns for two contradictory reasons: First, it is abstract, offering little guidance on practical issues such as what materials to employ or how to organize class activities. Second, a worked-out example such as that at PACHS, seems so embedded in its cultural-historical circumstances that it is hard to translate its insights to another setting. Dewey recognizes those contradictory pulls, examines them in detail, but offers no easy resolution. Instead, his theory, augmented by the analysis here, reminds us that supporting learning though inquiry means finding situation-specific ways to enhance *relevance* (relation of activities to learners' experiences), *participation* (mode of learners' engagement), and *significance* (impact on learners' individual and communal life). PACHS shows us a proof of concept, one way to achieve that for a particular group of students. In doing so, it reminds us that good teaching will always require creative inquiry itself.

References

Antrop-Gonzalez, R. (2003). This school is my sanctuary: The Pedro Albizu Campos Alternative High School. *Centro Journal, 15*(2), 232–255.

Armstrong, D. (2000). A survey of community gardens in upstate New York: Implications for health promotion and community development. *Health & Place, 6*(4), 319–327. https://doi.org/10.1016/S1353-8292(00)00013-7.

Benson, L., Harkavy, I., & Puckett, J. (2007). *Dewey's dream: universities and democracies in an age of education reform*. Philadelphia: Temple University Press.

Berry, P. W., & Cavallaro, A. J. (2014). Sustaining narratives of hope: Literacy, multimodality, and the Dr. Pedro Albizu Campos High School. *English Education, 46*(4), 279–299.

Bouillion, L. M., & Gomez, L. M. (2001). Connecting school and community with science learning: Real world problems and school–community partnerships as contextual scaffolds. *Journal of Research in Science Teaching, 38*(8), 878–898. https://doi.org/10.1002/tea.1037.

Boostrom, R. (2016). The peculiar status of democracy and education. *Journal of Curriculum Studies, 48*(1), 4–22. https://doi.org/10.1080/00220272.2014.962100.

Bruce, B. C. (2008). From Hull House to Paseo Boricua: The theory of practice of community inquiry. In B. Dicher & A. Luduşan (Eds.), *Philosophy of pragmatism: Salient inquiries* (pp. 181–198). Cluj-Napoca: European Studies Foundation Publishing House.

Bruce, B. C., & Bloch, N. (2013). Pragmatism and community inquiry: A case study of community-based learning. *Education and Culture, 29*(1), 27–45.

Burr, K. L. (2001). Building for hope: Progressive service learning enhances education. *Journal of Industrial Teacher Education, 38*(4), 84–94.

Cunha, M. V. D. (2016). We, John Dewey's audience of today. *Journal of Curriculum Studies, 48*(1), 23–35. https://doi.org/10.1080/00220272.2014.1003604.

Corrigan, M. P. (2011). Growing what you eat: Developing community gardens in Baltimore, Maryland. *Applied Geography, 31*(4), 1232–1241. https://doi.org/10.1016/j.apgeog.2011.01.017.

Cummings, C. K. (2000). John Dewey and the rebuilding of urban community: Engaging undergraduates as neighbourhood organizers. *Michigan Journal of Community Service Learning, 7*(1), 97–109.
Dewey, J. (1900/1976). The school and society. In J. A. Boydston (Ed.), *John Dewey: The middle works, 1899–1924* (Vol. 1, pp. 1–112). Carbondale: Southern Illinois University Press.
Dewey, J. (1902/1976). The child and the curriculum. In J. A. Boydston (Ed.), *John Dewey: The middle works, 1899–1924* (Vol. 2, pp. 271–291). Carbondale: Southern Illinois University Press.
Dewey, J. (1916/1980). Democracy and education. In J. A. Boydston (Ed.), *John Dewey: The middle works, 1899–1924* (Vol. 9). Carbondale: Southern Illinois University Press. (Reprinted from: 2008).
Dewey, J. (1938a). Experience and education. In J. A. Boydston (Ed.), *John Dewey: The later works, 1925–1953* (Vol. 13, pp. 1–62). Carbondale: Southern Illinois University Press. (Reprinted from: 1997).
Dewey, J. (1938b). Logic: The theory of inquiry. In J. A. Boydston (Ed.), *John Dewey: The later works, 1925–1953* (Vol. 12). Carbondale: Southern Illinois University Press.
Eisenhart, M., Finkel, E., & Marion, S. (1996). Creating conditions for scientific literacy: A reexamination. *American Educational Research Journal, 33*(2), 261–295.
Enfield, R. P. (2001). *Connections between 4-H and John Dewey's philosophy of education*. Davis: University of California. Retrieved from http://cyd.ucdavis.edu/publications/pubs/focus/pdf/FO01V7N1.pdf.
Flores-González, N., Rodriguez, M., & Rodriguez-Muniz, M. (2006). From hip-hop to humanization: Batey Urbano as a space for Latino youth culture and community action. In S. Ginwright, P. Noguera, & J. Cammorota (Eds.), *Beyond resistance! Youth activism and community change* (pp. 175–196). New York: Routledge.
Fusco, D. (2001). Creating relevant science through urban planning and gardening. *Journal of Research in Science Teaching, 38*(8), 860–877.
Gale, R. M. (2006). The problem of ineffability in Dewey's Theory of Inquiry. *The Southern Journal of Philosophy, 44*(1), 75–90. https://doi.org/10.1111/j.2041-6962.2006.tb00004.x.
Giles, D. E. J., & Eyler, J. (1994). The theoretical roots of service learning in John Dewey: Toward a theory of service learning. *Michigan Journal of Community Service Learning, 1*(1), 77–85.
Gruenewald, D. A. (2003). The best of both worlds: A critical pedagogy of place. *Educational Researcher, 32*(4), 3–12. https://doi.org/10.3102/0013189x032004003.
Johnson, L. R. (2005). *History in our hands: Identity development, cultural ideologies of motherhood, and the critical practice of family literacy in Puerto Rican Chicago.* (Ph. D dissertation), University of California, Berkeley.
Kahne, J., & Westheimer, J. (1996). In the service of what? *Phi Delta Kappan, 77*(9), 592–599.
Krasny, M. E., & Tidball, K. G. (2009). Applying a resilience systems framework to urban environmental education. *Environmental Education Research, 15*(4), 465–482. https://doi.org/10.1080/13504620903003290.
Mayhew, K. C., & Edwards, A. C. (1936/1965). *The Dewey School: The laboratory school of the University of Chicago 1896–1903*. New York: Atherton Press.
Moll, L. C., Amanti, C., Neff, D., & Gonzales, N. (1992). Funds of knowledge for teaching: Using a qualitative approach to connect homes and classrooms. *Theory into Practice, 31*(2), 132–141.
Mueller, M. P., & Tippins, D. J. (2012). Citizen science, ecojustice, and science education: Rethinking an education from nowhere. In B. J. Fraser, K. G. Tobin, & C. J. McRobbie (Eds.), *Second international handbook of science education* (Vol. 2, pp. 865–882). Dordrecht: Springer.
Oakes, J., & Rogers, J. (2006). *Learning power: Organizing for education and justice*. New York: Teachers College Press.
Osborne, M. D. (1998). Responsive science pedagogy in a democracy: Dangerous teaching. *Theory into Practice, 37*(4), 289–295.
Ramos-Zayas, A. Y. (1998). Nationalist ideologies, neighbourhood-based activism, and educational spaces in Puerto Rican Chicago. *Harvard Educational Review, 68*(2), 164–192.

Roth, W.-M. (2007). Identity in scientific literacy: Emotional-volitional and ethico-moral dimensions. In W. M. Roth & K. Tobin (Eds.), *Science, learning, identity: Sociocultural and cultural-historical perspectives* (pp. 153–184). Rotterdam: Sense Publishers.

Saltmarsh, J. (1996). Education for critical citizenship: John Dewey's contribution to the pedagogy of community service learning. *Michigan Journal of Community Service Learning, 3*(1), 13–21.

Tanner, L. N. (1997). *Dewey's laboratory school: Lessons for today*. New York: Teachers College Press.

Torrise, M. (2010). Role of the library media specialist in greening the curriculum: A community-based approach to teaching 21st century skills outside of the school library through the practice of urban agriculture. *Library Media Connection, 28*(4), 18–20.

Weissman, E. (2015). Entrepreneurial endeavours: (Re)producing neolibralization through urban agriculture youth programming in Brooklyn, New York. *Environmental Education Research, 21*(3), 351–364. https://doi.org/10.1080/13504622.2014.993931.

Won, M. (2009). *Issues in inquiry-based science education seen through Dewey's theory of inquiry.* (Ph.D. dissertation), University of Illinois at Urbana-Champaign, Champaign, IL.

Chapter 3
Forging Research Pathways to Sustainable Farms and Food Systems with an Interdisciplinary Evaluative Framework for Urban Agriculture

Helena K. Farrell

Abstract Urban agriculture manifests in a myriad of ways across the globe and is at the intersection of urgent contemporary issues, including social justice, environmental sustainability, community development, and climate change adaptation and resilience. The study of urban agriculture is inherently interdisciplinary as urban farms are a nexus where technological and design innovation, visionary urban land use and policies, and community-based social and economic organizing take place. New urban agriculture research and teaching strategies are needed that help to differentiate designs and practices, evaluate a wide spectrum of impacts, and prepare students to understand and participate in urban agriculture work that is linked to current events and challenges. This chapter describes how a custom urban agriculture research method was used by students in an undergraduate course at the University of Massachusetts, Amherst to forge research pathways into topics that were compelling to them and that represent some of the most urgent sustainability challenges and existential threats currently facing communities worldwide. By providing the method, tools and coaching for students to develop seldom taught interdisciplinary skills, this course helped to prepare students to succeed in urban agriculture and sustainability professions and to demonstrate the vital role of urban agriculture in creating sustainable, equitable, and resilient communities in the context of twenty-first century challenges.

H. K. Farrell (✉)
Former Lecturer and Visiting Scholar, University of Massachusetts, Amherst, MA, USA

Landscape Architecture & Regional Planning, Stockbridge School of Agriculture, Amherst, MA, USA

I. DeCoito et al. (eds.), *Teaching and Learning in Urban Agricultural Community Contexts*, Urban Agriculture,
https://doi.org/10.1007/978-3-030-72888-5_3

Keywords Urban agriculture · Case study research · Agroecology · Farm system design · Sustainability education · Interdisciplinary skills · Systems thinking · Teaching strategies · Student engagement · Experiential learning

3.1 Forging Research Pathways to Sustainable Farms and Food Systems with an Interdisciplinary Evaluative Framework for Urban Agriculture

The increasing demand for urban agriculture education (Fritsch 2013), and interdisciplinary environmental and sustainability education (Vincent et al. 2015) parallels the worldwide trend that is focusing on urban agriculture development in response to inequities and shortfalls of the industrialized food system (Mujčinović et al. 2019). These trends highlight the urgent need for new approaches to education, practice and policy that effectively position urban agriculture to respond with the sustainable social, economic and environmental benefits that it can provide (Mujčinović et al. 2019) to communities threatened by food insecurity, industrial farming and food system vulnerabilities, and the rapid and dramatic changes in the environment and climate (McClintock 2014; Nugent 1999; Pearson 2010).

A majority of students consider environmental academics when applying to schools, and institutions of higher education in the U.S. have responded by quickly establishing new programs, while existing programs have struggled to keep up with student demand (Vincent et al. 2015). Students are interested in learning about urban agriculture, an important subject within sustainability education, as an alternative to disciplinary education and careers (Martens 2005) and as an alternative to traditional agricultural education which emphasizes behavioral skills needed in traditional agricultural professions. The study of contemporary urban agriculture in courses based on systems thinking and understanding social, economic, and environmental dimensions of urban agriculture, emphasizes conceptual skills and learning about other important disciplines involved in successfully implementing urban agriculture in a way that prepares students for careers linked to sustainable agriculture, community organizing and resilience, food access and social justice, food system transformation, biodiversity and natural resources protection, and climate change adaptation. New teaching strategies, customized research tools, and coaching are needed to meet current urban agriculture education demands (McClintock 2014; Mougeot 2000; Parr et al. 2007). Additionally, interdisciplinarity is central to students excelling as professionals and effectively addressing critical sustainability challenges through urban agriculture (Hilimire et al. 2014; Vincent et al. 2015).

Agroecology uses ecology to design and manage sustainable food systems. By considering social, economic, and environmental impacts of agriculture, it offers integrated solutions for sustainably feeding the growing world population and creating equitable farming and food systems while protecting natural resources (Gliessman 2020). Agroecology has been recognized as having an important role in reorientating agricultural systems toward highly productive and sustainable modes of production that contribute to the progressive realization of the human right to adequate food (De Schutter 2010), and agroecological systems approaches for sustainable food system transformation have been increasingly developed, implemented, documented, and embraced in the past decade (Gliessman 2020). Agroecology's systems approach, interdisciplinary and action-oriented nature, and critical focus on social, environmental, economic outcomes as well as agricultural sustainability provide the ideal framework for teaching contemporary urban agriculture research and practice.

The author describes the development of an interdisciplinary urban agricultural framework and research method based on the philosophies and assumptions of Agroecology and how the framework and method were used in an undergraduate course titled, *Urban Agriculture: Innovative Farming Systems for the twenty-first Century*, at The University of Massachusetts, Amherst (UMass). The course was part of the Sustainable Food and Farming Program at UMass, Amherst in which students earned credit towards a Certificate, Associate's or Bachelor's Degree in *Sustainable Food and Farming.* Students in other UMass departments and degree programs and from the other Five Colleges: Smith, Amherst, Mount Holyoke, and Hampshire Colleges, also enrolled in the course. Between 2012 and 2016, *Urban Agriculture: Innovative Farming Systems for the twenty-first Century* was taught four times online, twice in-person, and was completed by ninety students. There were no prerequisites or specific prior skills or knowledge required for registered college students to enroll.

The course presented a pedagogical model along with the rationale behind it that equipped college students to investigate contemporary urban farm systems as the primary focus of urban agriculture, including its critical social, economic, and environmental dimensions. Examples of student research representing a range of topics are presented, including key findings and sustainability implications. The challenges of applying this approach in the context of this course are also discussed.

3.2 Origination: The Need for a Custom Framework and Research Method

The need for better articulation and more accurate distinctions between diverse urban farm systems in order to comparatively analyze and assess different designs for a Landscape Architecture Masters Project, *Illuminating Urban Agriculture: A New Framework for Understanding Complexity* (Farrell 2011) provided the impetus for developing the comprehensive evaluative framework and research method for urban agriculture. Integrating accepted standards of measurement, referred to in this chapter as "value metrics", from relevant disciplines into a comprehensive evaluative framework was an essential task needed to support the assessment of urban farms for their sustainability impacts and critical outcomes. The comprehensive evaluative framework and assessment method have been applied internationally in the United Kingdom (Laine 2011) and Ecuador (Palacios and Elizabeth 2013). It also provided value metrics and assessment criteria to a study commissioned by the Local Food Research and Development Division of USDA Agricultural Marketing Service, which assessed the economic viability of U.S. urban farms through an interdisciplinary case study approach (Rangarajan and Riordan 2019).

3.3 Built-in Interdisciplinarity: The Comprehensive Evaluative Framework

3.3.1 Value Metrics for Farm System Analysis

Prior to April Philips' 2013 publication of *Designing Urban Agriculture: A Complete Guide to the Planning, Design, Construction, Maintenance and Management of Edible Landscapes* (Philips 2013), there was minimal explicit analysis of designed urban agricultural systems within the academic literature. Previous research was focused largely on understanding a small number of urban agriculture's social and economic impacts. The comprehensive evaluative framework outlined in this chapter, however, expands to include additional social and economic benefits, ecological benefits, as well as gross agricultural yield as additional, critical outcomes of urban farm systems.

Value metrics from a range of disciplines important for the successful implementation of urban agriculture and the full accounting of its value across social, environmental, economic, and agricultural interests are built into the framework, including contemporary urban agriculture (Nugent 2000), conventional agriculture, permaculture and forest gardening (Jacke and Toensmeier 2005), agroecosystem analysis, landscape architecture, urban planning, and neoclassical and informal economics.

A value metric is a standard unit or system for measuring relevant data concerning the farm system and may take different forms: numeric value, qualitative description, illustration, a trend, or the presence or absence of something. Methods for collecting data include site visits and site analysis, interviews, and information gathering from the library and web. A survey sheet was created for collecting farm system data.

Value metrics for farm system analysis are mapped out within their respective dimension of the farm system: "System Design", "System Dynamics", "Outcome and Benefits", as illustrated in Fig. 3.1. Outcomes and Benefits are linked to their foundation in System Design via a critical set of System Dynamics. Displaying these three dimensions of urban farm systems and their associated value metrics as visually interconnected to one another helps to make the comprehensive evaluative framework conducive to systems thinking. Interdisciplinarity is built into this research method through the synthesis of these metrics and their interrelatedness in the framework. The premise for this systems approach is that designing optimally performing farm systems begins with understanding the web of influences in which the farming system is embedded, and that outcomes and benefits are the result of farm system design and dynamics. The framework and matrices (explained next) are visual, analytical tools for systematically exploring individual aspects of a farm system with an eye on the whole system and potential impacts beyond.

3.3.2 Assessment Matrices

Once the data has been gathered, twelve assessment criteria are used to evaluate farm system performance. Data is cross-referenced with assessment criteria using an assessment matrix for each farm system dimension: System Design (Fig. 3.4), System Dynamics (Fig. 3.5), and Outcomes and Benefits (Fig. 3.6). The matrices make it possible to evaluate the data collected, to rank performance, and to determine what aspects of the farm system warrant further investigation. This method of examination reveals capacities, strengths, weaknesses, and sustainability impacts.

Like the value metrics, assessment criteria are a synthesis of contemporary best practices articulated by different disciplines involved in sustainability, including environmental science, sustainable agriculture, urban planning, landscape architecture, economics, and permaculture. Evaluating farm systems at site-scale according to sustainability goals and best practices is important for verifying urban agriculture's role in addressing key sustainability challenges locally and globally and for urban agriculture practitioners to be effective at reaching their goals. Assessment criteria included:

- Turns problems into solutions
- Observes, interacts, responds
- Values and integrates diversity
- Produces abundant, complimentary yields
- Renews water, energy, fertility and inputs
- Regenerates organic matter and topsoil
- Captures waste from the urban waste stream
- Produces no pollution or waste
- Reduces food miles
- Fosters neighborhood safety and community cohesion
- Fosters access to food, education and employment
- Fosters formal or informal economy

While this research method was tailored for urban farm systems, it may be applied to sustainable farming and food systems projects working to maximize particular social, ecological, or economic outcomes in addition to agricultural productivity through systems thinking approaches and designed complexity. Due to the framework's comprehensive, interdisciplinary, and adaptable nature, virtually any case study or topic related to urban agriculture can be investigated with this research process. The research tools – the evaluative framework and assessment

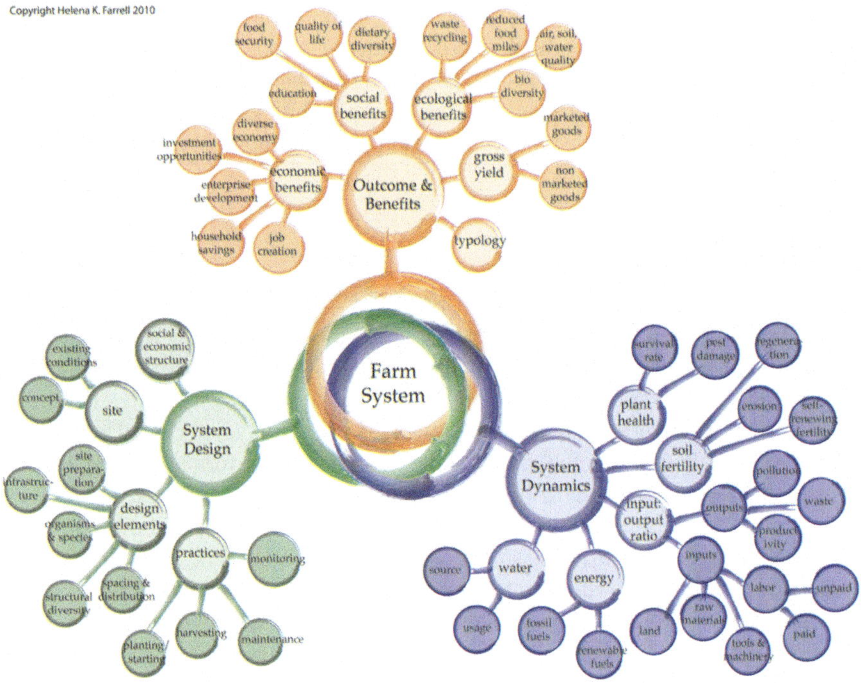

Fig. 3.1 Comprehensive evaluative framework

matrices – can be customized by switching out value metrics or assessment criteria to support different inquiries. This process can support research that is extremely specific, or broad and holistic; problem-oriented or inquiry-oriented; focused on skill-building, and technical problem-solving, or aimed to address community issues; provide service learning and non-traditional collaboration opportunities; or generate new scholarship for the academic community. If used for teaching, the framework, matrices, and research methods should support learning objectives that students can achieve and suit the learning environment where they are used.

3.4 Interdisciplinary and Experiential Learning through Urban Agriculture Research: From Concept to Practice

3.4.1 Course Learning Objectives

The five learning objectives of the course were: (1) implement agroecology as an interdisciplinary approach to learning about urban agriculture; (2) identify and describe the origins and driving forces of urban agriculture; (3) explain best practices, innovative production methods, and agricultural careers, and why urban

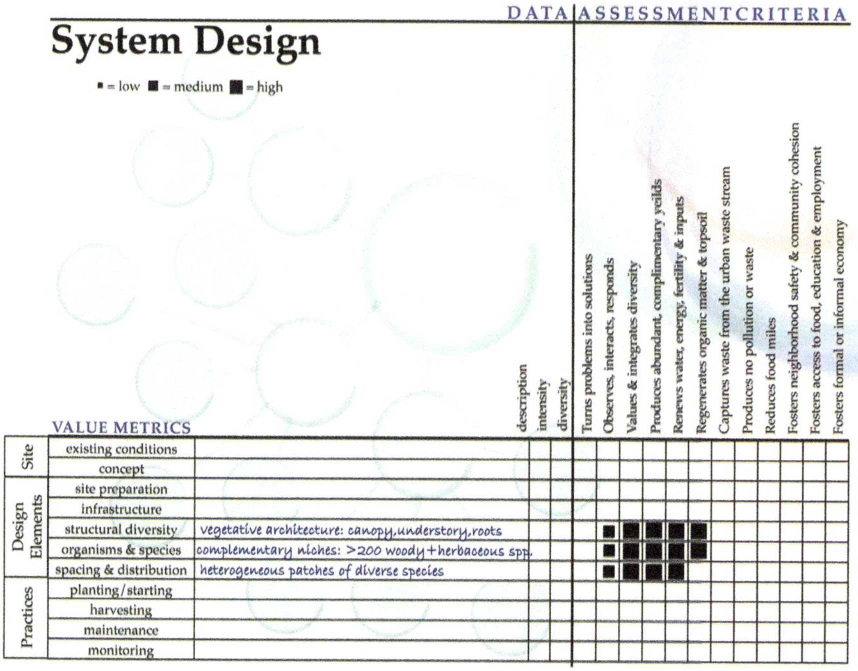

Fig. 3.4 System design matrix for the Holyoke Edible Forest Garden

agriculture is needed in the context of current events and contemporary issues; (4) situate and analyze urban farms as complicated, dynamic systems with a range of outcomes and benefits; and, (5) apply critical thinking, scholarly research, and communication skills.

3.4.2 *Teaching Strategies and Student Engagement to Support Learning Objectives*

Like a *Choose Your Own Adventure*, students used the comprehensive evaluative framework (Fig. 3.1) to focus and frame their research; to forge research pathways into sustainable farming and food systems topics with urban farm systems as the specific focus area. The evaluative framework provided a road map for navigating interrelationships between farm system designs, dynamics, outcomes and benefits, and for understanding how multiple disciplines come together in real life.

Individual metrics provided entry points from which a path of investigation could be charted based on the students' research questions. Like a map, the framework was a visual key for choosing where to start, how far to explore, and which connections to make, and not to make, along the way. Discovering the interdisciplinarity of

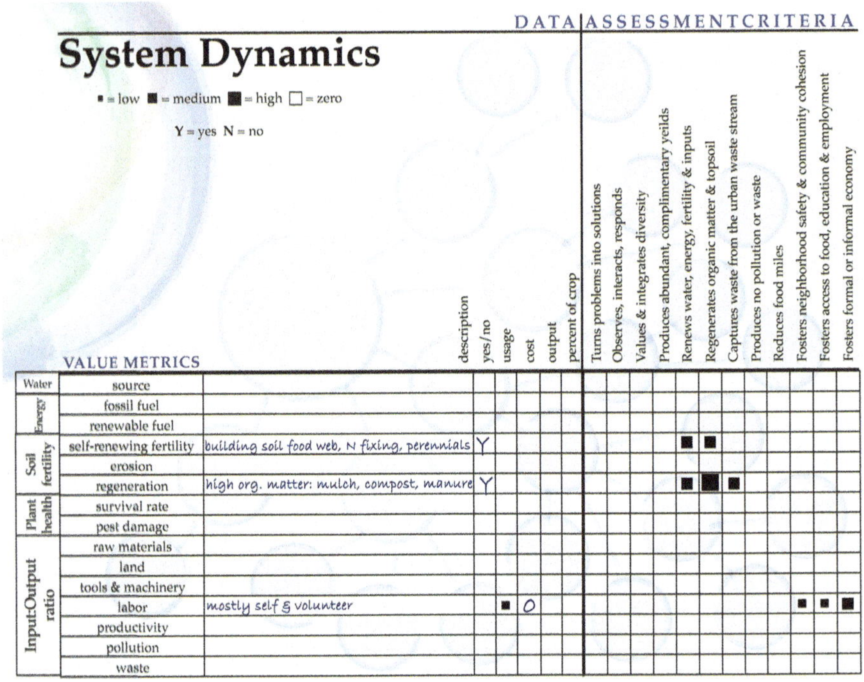

Fig. 3.5 System dynamics matrix for the Holyoke edible forest garden

urban agriculture was an emergent process, in that, once research questions had been focused and framed, students subsequently encountered topics from other disciplines. This provided the impetus to become familiar with and draw from those disciplines in order to understand and address their research questions. This teaching strategy, when combined with instructor support and oversight, nurtured students' ability to work from multiple theoretical perspectives and helped them build confidence and stamina for taking on leadership roles in the future.

Students were instructed on how to ask critical questions and completed exercises in which they were required to practice observation as the critical basis for inquiry (Hilimire et al. 2014). Getting students to ask questions and drawing upon their interests and life experiences fostered engagement, which is especially important when critical concepts of agriculture or other important disciplines are unfamiliar. Pursuing how-and-why questions is the heart of agricultural science education and the nexus where biological, physical, and social sciences interconnect within urban agriculture (Martens 2005).

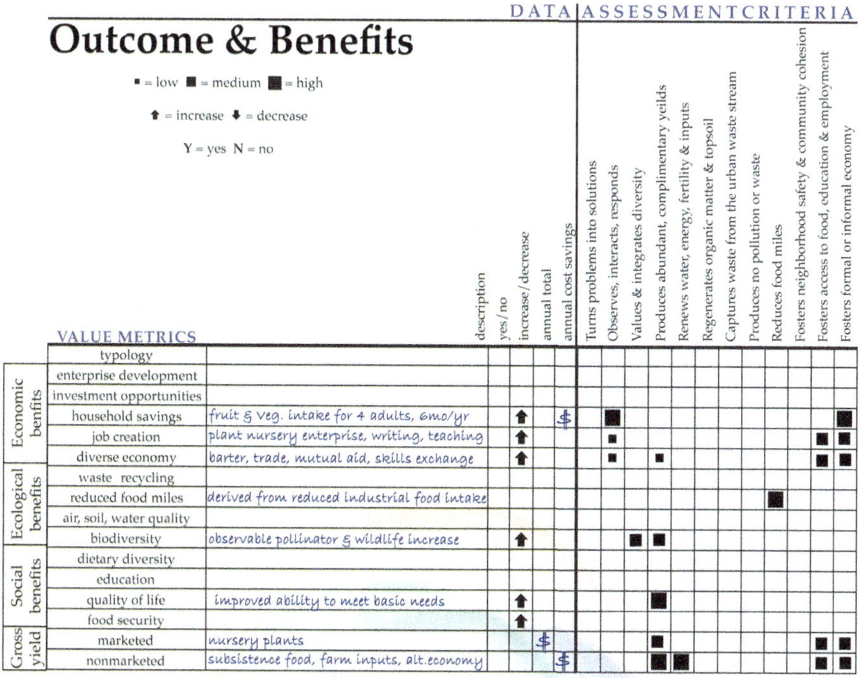

Fig. 3.6 Outcomes and benefits matrix for the Holyoke Edible Forest Garden

3.4.3 *The Research Process*

Working individually in the on-line version of the course, and in small groups in the in-person version, students were instructed to follow a three-part case study or topic research process: analysis; assessment; conclusion. Value metrics were used for analysis, assessment criteria were used for assessment, and the conclusion was used to present research findings and next steps. Students charted their research pathway by analyzing value metrics from the comprehensive evaluative framework and assessing their data findings with the assessment matrices. Figure 3.2 shows examples of students' research pathways. Figures 3.4, 3.5 and 3.6 show examples of the assessment matrices. Students ranked the performance of value metrics that they analyzed, determined successes, strengths, weaknesses, best practices and key lessons, and identified aspects of the research that warranted further investigation and by what means. Students concluded their research with key findings, design recommendations, or next steps. Some students used their research to develop a grant proposal, business plan, farm system design, or advocacy initiative.

Case studies provided an opportunity for students interested in a particular kind of farm system to learn from a specific precedent whereas research topics allowed students to investigate a subject or theme broadly. Typically, both were chosen based on what students had seen or experienced and what they wanted to know

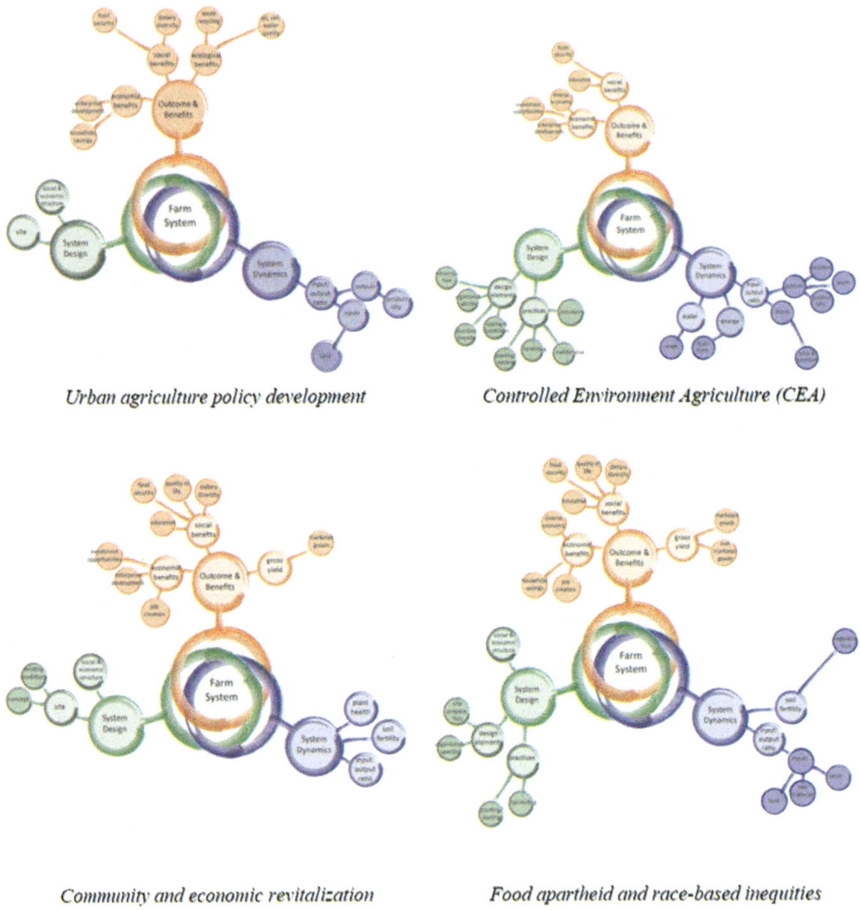

Fig. 3.2 Examples of research pathways associated with students' interdisciplinary research

more about in order to help them make informed career choices. For example, one student studied an urban cut flower farm and retail florist in Philadelphia asking, "Can the farm system be economically viable at the small scale of an urban neighborhood? Can the system be designed to save on chemical inputs, packaging and transportation, providing a comparative advantage over the existing commercial flower industry?" The student used her case study research to answer these critical questions and develop an understanding of the business model she was interested in pursuing after college. Other topics included, addressing food deserts and race-based inequities in the food system, assessing the need for urban agriculture policy and advocacy, fostering biodiversity by creating wildlife habitat on urban farms, and municipal programs for composting urban food waste. One student studied the artistry and graphic design of WWII victory garden propaganda to understand the role and significance of media in creating a collective vision to sustain a social and political movement around community food systems. She sought to apply what she learned from that historic effort to the current local food and urban farming movement.

At the end of each course, students shared their research with the rest of the class and sometimes also with the practitioners and organizations with whom they had established a relationship. The in-person course dedicated several class periods to student presentations, which fostered peer-to-peer learning and a sense of contributing collectively to a new and growing body of knowledge in contemporary urban agriculture research and scholarship. Table 3.1 shows a variety of research conducted in the course, disciplines that were relevant to the investigation, and actual professional opportunities that resulted from students' research during the course. Figure 3.2 shows the research pathways associated with each example listed in Table 3.1. A descriptive summary of one case study follows Table 3.2. More examples of student work can be found in Table 3.2.

Table 3.1 Examples of students' interdisciplinary research

Case study or research topic	Research question	Relevant disciplines	Professional opportunity
Urban agriculture policy development in Providence, RI	How can state and local agricultural laws and policies foster a sustainable, food secure city in Providence, RI?	Public policy, City planning, public health	Community policy and advocacy
Controlled Environment Agriculture (CEA)	How does CEA measure up as a highly efficient and productive form of agriculture when assessed across five parameters: Community development, food security, low input, high output, and education?	Business management, education, community development	Employment at local hydroponics farming operation
Community and economic revitalization through urban agriculture	How can communities and local urban agriculture businesses work together to improve quality of life through sustainable urban agriculture enterprise development?	Economic development, City planning, Urban ecology	Internship with City of Lowell, MA.
Food apartheid and race-based inequities in the food system	What is the role of urban agriculture to address disparities in food access, nutrition and health in underserved urban communities?	Political economy, social justice, public health, community Food systems	

3.4.4 The Holyoke Edible Forest Garden: One Case Study Research Summary

This section provides a summary of how one student applied the research method to analyze and assess an urban farm case study, develop key findings that address their research questions, and identify knowledge gaps and opportunities for future investigation. This and other examples of student research are shown in Table 3.2.

Through a site visit to the Holyoke Edible Forest Garden and interviews with the practitioners, the student analyzed and assessed an urban farm case study in the form of a mature, permaculture-style, edible forest garden cultivated by residents of a two-family home in an urban neighborhood in the City of Holyoke, Massachusetts. This research identified important findings by using value metrics (shown in bold text) as the focus of data gathering, and then using the assessment matrices to assess that data according to sustainability criteria (Figs. 3.4, 3.5 and 3.6). The research pathway for this case study is illustrated in Fig. 3.3.

Analysis and assessment of the edible forest garden's system design focused on **structural diversity, spacing and distribution,** and **organisms and species** (Fig. 3.4). These metrics were chosen to target the most pertinent data for assessing the high productivity, multi-functionality, low labor and inputs, and diverse yields achieved by the edible forest garden within the constrained 1/10 acre **site**. The student found that **structural diversity**, which makes use of vertical space for production, was achieved using the natural vegetative architecture of the plants: fruit tree canopy, herbaceous understory, ground layer, and root zone. The **spacing and distribution** had plants growing very close together in efficient patterns maximizing production per square foot. **Organisms and species** growing in the edible forest garden required careful selection to match biological suitability with the growing conditions, which were highly dense, diverse, and productive. The discussion of these three metrics raised further research questions relevant to contemporary sustainability and community resilience challenges, including: how much carbon can high-density perennial agriculture landscapes, such as edible forest gardens, sequester from the atmosphere and store in the soil; how are soil nutrients and moisture sustained and renewed in perennial agriculture compared to tillage agriculture, and how does the nutritional value of yields compare; and how will ecological niches be designed for resilience to increasing average temperatures and extreme weather events, and what perennial plant adaptations and characteristics will be selected for climate change resilience?

Analysis and assessment of the edible forest garden's system dynamics (Fig. 3.5) revealed a unique approach to achieving a range of desirable outcomes by establishing beneficial ecological relationships that reduce human labor inputs. **Labor** calculations were low because most of the plants were perennial, and once the garden was designed and installed, the majority of on-going labor was to harvest food. As it matured, the forest garden established an extensive soil-food-web in which fungi and microbes would break down organic matter into nutrients for plant uptake achieving **self-renewing fertility**. **Soil regeneration** occurred largely without

Table 3.2 Examples of student research, key findings, and major products

Case study or topic	Key findings	Skills	Self-directed or collaborative	Modes of investigation	Major products
Urban agriculture policy development in Providence, RI	Analysis of current agricultural law reveals opportunities to craft urban agriculture policy for food security and economic development at the commercial, community and residential levels, and give the City of Providence a legal voice in statewide policy creation.	Case study analysis and assessment, critical thinking and problem-solving across multiple scales and disciplines	Self-directed	Community engagement, Student-centered, literature review	Food security model for the City of Providence, RI., comprehensive urban agriculture policy recommendations
GrowNYC, NY Added value farms, Brooklyn, NY YouthGROW, Worcester, MA	Case studies show how urban agriculture and community food systems projects can offer social, economic, and ecological strategies in food deserts to address the scarcity of fresh, healthy food, poor nutrition, poverty, and disease.	Critical observation of topic area, case study assessment	Peer-to-peer group project	Case studies, literature review	Community development strategies, Sustainable food system business concepts and best practices

(continued)

Table 3.2 (continued)

Case study or topic	Key findings	Skills	Self-directed or collaborative	Modes of investigation	Major products
Utilizing a food therapy approach through urban farming to help individuals with eating disorders	The experience of food cultivation can support one's recovery from an eating disorder. At the same time, hospitals can provide opportunities for social engagement and environmental stewardship through the creation of gardens for healing and food production.	Critical observation of topic area, professional goals articulation, career assessment	Self-directed	Student-centered, career-path planning	Business plan: *Proposal for a Food Therapy Program in New York City Hospitals*
Windham community garden, ME & Alan day Community Garden, ME	Community gardens dependent on public water infrastructure can safeguard their supply, reduce energy usage and costs by designing and integrating rain harvesting, storage and distribution systems to supplement water consumption.	Case study analysis and assessment	Self-directed	Case studies, community engagement	Design recommendations, water harvesting design details

(continued)

Table 3.2 (continued)

Case study or topic	Key findings	Skills	Self-directed or collaborative	Modes of investigation	Major products
Racism and inequity awareness assessment	Increased awareness is needed to ensure that urban agriculture combats rather than perpetuates racial and socioeconomic injustices. Projects must address oppressive structures underlying existing social inequities as well as offer the benefits communities need.	Critical observation of topic area, critical thinking and problem-solving across multiple scales and disciplines	Peer-to-peer group project	Case studies, literature review	Next steps for research, education and policy

(continued)

Table 3.2 (continued)

Case study or topic	Key findings	Skills	Self-directed or collaborative	Modes of investigation	Major products
Green roofs and rooftop farms	The benefits of green roofs and rooftop farms vary: Some are viable, food producing businesses characterized by innovation and job creation. Some offer ecosystem services such as storm water mitigation and improved building efficiency. Others provide education, recreation and community engagement. Rather than follow a universal model, green roofs and rooftop farms must be customized to suit the inputs, outputs, and benefits needed where they exist. In this way, the three E's of sustainability (equity, environment, economy) are achieved as optimally as possible, and as often as possible.	Site analysis and assessment, design thinking, contemporary observation of topic area	Peer-to-peer group project	Case studies, literature review	Sustainable business concepts and best practices

(continued)

Table 3.2 (continued)

Case study or topic	Key findings	Skills	Self-directed or collaborative	Modes of investigation	Major products
Cooperative food retailers for social justice	Urban areas with limited access to fresh food should be the geographical target of cooperative food retailers providing education and civic engagement around issues such as transportation, environmental justice, public health, and economic empowerment.	Critical observation of topic area, critical thinking and problem-solving across multiple scales and disciplines	Self-directed	Community engagement, student-centered, career-oriented	Advocacy and policy recommendations, sustainable business concepts and best practices
Ecological residential site design	Implementing ecological landscapes and infrastructure at residential scale can reduce costs, increase dietary diversity and quality of life, and foster social and environmental engagement of residents.	Site analysis and assessment, design thinking and problem-solving, design drawing	Self-directed	Student-centered	Conceptual landscape design illustrations

human intervention, as leafy debris from the densely grown plants would fall and decompose on the soil surface. The investigation of these two metrics represents the interdisciplinarity of urban agriculture research in that the student reached an appropriate juncture to draw from the discipline of plant and soil science by using soil tests to measure nitrogen, phosphorous, potassium, trace minerals, and soil organic matter in order to more accurately assess the self-renewing fertility and soil regeneration occurring in the forest garden.

Analysis and assessment of the edible forest garden's outcomes and benefits (Fig. 3.6) showed that the forest garden provided substantial **household savings**

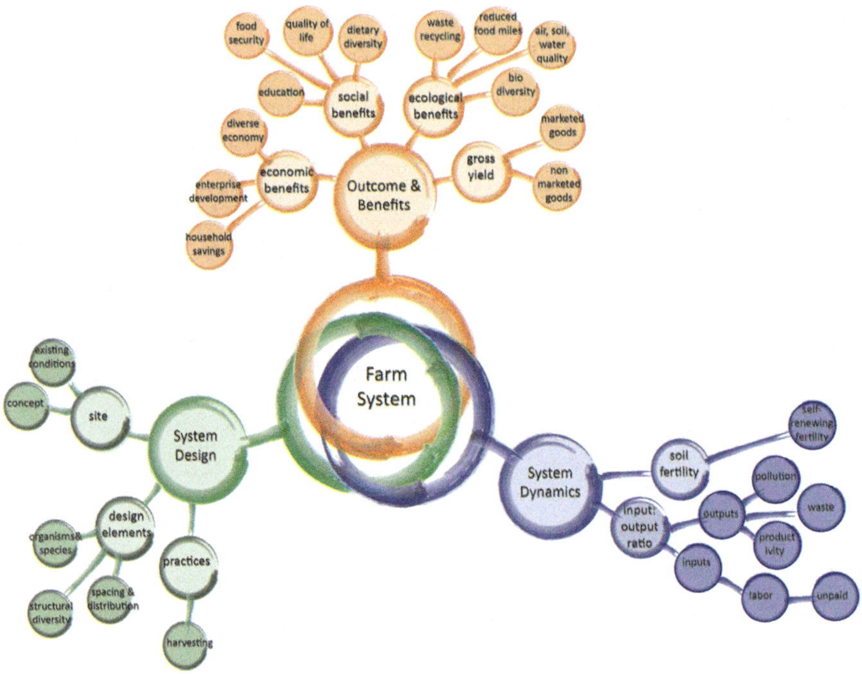

Fig. 3.3 The research pathway of the Holyoke Edible Forest Garden case study

(approximately 75–95%) on fruit and vegetable purchases for two families for 6 months of the year, had resulted in an observable increase in wildlife habitat and **biodiversity** on site, specifically pollinator insects as well as birds and generalist wildlife, and was fostering **enterprise development** through the formal sale of **marketed goods, job creation** through teaching and writing opportunities for the practitioners, and **diverse economy** in the form of barter, trade, mutual aid, and the informal exchange of skills and services among neighbors and friends.

This case study illustrates that following a targeted research pathway within a holistic framework can identify important relationships between system design, system dynamics, and outcomes and benefits, which influence the farm system's overall character, its agricultural performance, its benefit to the community, and its impact on important global challenges related to sustainability and resilience. For example, the edible forest garden was assessed as having a "high" rating for reducing food miles because it significantly reduced household consumption of commercially available food from industrialized farm systems. This outcome originated in the design and dynamics of the farm system, yet the impact, assessed using sustainability criteria in the matrices, addressed issues with the global food system.

Methodological issues encountered in this study included the challenge of calculating reduced food miles as an ecological benefit related to the reduction of commercially sourced fruits and vegetables for the two families. Knowledge gaps included the gross yield of all marketed and non-marketed goods produced by the forest garden, which was unknown simply because the residents had not been keeping track of it all. The student could have devised a system to measure and record this data, but they couldn't wait a year for that data to be generated, nor could they add the collection of this data to the practitioners' task list.

See Table 3.2 for more examples of student research, including key findings, skills practiced, collaborative or self-directed learning examples, modes of investigation used, and major products of the research.

3.4.5 The Value of Experience

The combination of hands-on experience with classroom instruction provides students with a more complete foundation for understanding and engaging in the culture, science, scholarship, and practice of urban agriculture (Fritsch 2013; Hilimire et al. 2014; Parr et al. 2007). With interdisciplinarity built into the research process, students could see how a range of expertise was interconnected and contributing to their learning, while practical, hands-on experience helped show how research can translate into action, how science can be applied in the field (Parr et al. 2007), and how evidence and experimentation from multiple disciplines can influence urban agriculture through farm system design, system dynamics, and outcomes and benefits.

In the course *Urban Agriculture: Innovative Farming Systems for the twenty-first Century*, students developed career-focused research pathways that aligned their learning in the course with their educational and professional goals. They also gained hands-on experience by conducting surveys and interviews with people in real-life to inform their research, by visiting farms and taking part in farm work, by volunteering in food systems activities and outreach, and by pursuing jobs and internships when those opportunities arose. The in-person version of the course featured field trips and volunteer opportunities at nearby urban farms. By utilizing this approach in the course, students were positioned to enact two roles, as participant and observer, which provides the basis for meaningful learning (Hilimire et al. 2014).

Occasionally, students had the opportunity to experience hands-on activities that physically shape and transform the landscape for the pursuit of food production, soil regeneration, waste recycling, and so on. Working with ones' hands as well as ones' mind provides meaningful educational experiences that are some of the most influential factors for choosing a career in agriculture (Esters and Bowen 2005).

3.4.6 Contemporary Urban Agriculture Information Gathering and Scholarship

The course instructor worked with the UMass library liaison to create a custom, urban agriculture subject guide within the UMass library's online databases and implored students to search the vast collections of scholarly journals on agriculture, science, technology, and sustainability topics for references to support their work. The in-person version of the course dedicated a full class period to a library database tutorial with the library liaison. The online version of the course provided database access and asynchronous tutorials on the course homepage. Students were required to gather and reference scholarly literature in their research papers to help develop cogent arguments to verify and expand upon assumptions they may have had, and to identify knowledge gaps. Practicing this traditional academic skill was an important complement to students' self-directed and empowered learning. Exploring the databases also helped to familiarize students with urban agriculture research and practice occurring globally, which tended to differ from current examples in the U.S.

The process of gathering information often involved refocusing and adjusting research questions due to knowledge gaps and methodological limitations, and students were instructed on how to document these limitations so that the need can be addressed by future researchers or problem-solvers. By practicing these scholarship skills and sharing their research with one another and with practitioners in the field, students made their own contributions to the body of knowledge of urban agriculture, which is constantly being developed (Hilimire et al. 2014). This research process and the teaching of it is an example that meets the need for interdisciplinary and applied scholarship in sustainable agriculture education (Parr et al. 2007).

3.4.7 Challenges to Contemporary Urban Agriculture Research at College-Level

Prioritizing scholarly literature in lieu of content from the internet was challenging due to a shortage of new research on urban agriculture from an interdisciplinary standpoint compared to the traditional, highly focused disciplinary research on urban agriculture that mainly appears in scholarly journals. It was also difficult to find accepted and accessible methods for calculating and assessing value metrics related to sustainability, which can be broad and difficult to measure. Establishing a foundation of knowledge on contemporary urban agriculture to support learning while simultaneously working to produce new knowledge through research represented a challenge for both the students and the instructor.

Students often found that the internet was a better resource than the university library databases for studying urban agriculture as a nexus of social, ecological, and economic issues and as a focus area for contemporary urban agriculture careers related to sustainability, social justice, community resilience, and transforming the food system, which were the topics that interested them. The instructor maintained the importance of referencing scholarly literature, highlighting that knowledge gaps represented opportunities for new research, tools, and methodologies. The instructor also noted that students' own first-hand information and research findings were all the more valuable in the context of scarce, new, interdisciplinary research, which helped to motivate students to overcome the sometimes logistically challenging task of collecting first-hand information through direct inquiries with practitioners in the field.

Applying this research method in the context of the course was inherently challenging due to its potential breadth and complexity. Equipping students to succeed at this approach required ongoing instructor oversight and support as students worked to define their research questions and forge their research pathways. Students had to be self-motivated to explore new territory and also willing to pursue and discuss open inquiries in which their findings might contradict their preconceived notions. The research process required sustained attention and stamina, and credible scholarship required maturity and follow-through; features which often needed to be developed in most students. Instructor oversight and support was important for keeping students engaged, as was simplifying their inquiries and research pathways to fit comfortably with their capacities and interests. Though the course description was forthright about its rigorous, investigative approach, some students enrolled who only wanted a cursory understanding of urban agriculture; more of a show-and-tell learning experience. These students tended to struggle more with engagement and resistance to the research requirements.

Students who invested in the research process, who came to understand urban farms as complex, adaptable systems with varying social, ecological and economic impacts, and who developed the ability to think critically about the issues and traverse academic or professional boundaries in pursuit of explanations and solutions, will have benefitted by becoming better prepared for careers in urban agriculture and sustainability after college. They also will have become better equipped to fully account for the value urban agriculture and to advocate successfully for it in the future by having conducted substantial and credible research from a comprehensive and interdisciplinary perspective.

The greatest challenge to urban agriculture research and sustainability education may be that despite high student demand, broad societal need, and support from science policy experts, the commitment of universities to fostering interdisciplinary research is still in question (Leahey et al. 2019). The majority of programs that do exist lack autonomy, resources, and institutional support compared to departments built around traditional academic disciplines. Staffing with adjunct instructors is one symptom of this problem, which leaves programs more vulnerable to attrition and the turnover of talented educators who are offered only part-time, short-term positions that are often low-paying, and without benefits. Adjunct positions are also more vulnerable to layoffs resulting from uncertainty in the educational services industry, for example as resulted from the COVID-19 pandemic (Burke 2020). Higher education must continue to rethink institutional structures to identify and implement objectives that improve the support, status, and inclusion of contemporary urban agriculture scholarship and sustainable food and farm systems education in academia.

3.5 Conclusion and Next Steps

New and forward-thinking educational strategies are needed to equip students with twenty-first century skills and problem-solving abilities so they will be able to take on the existential threats currently facing human society worldwide, including severe weather brought by climate change, food system vulnerabilities, pandemic-related supply chain disruptions, chronic hunger and food insecurity, economic and race-based health disparities stemming from lack of access to healthy food, wage inequity for food system workers, and farmland access disparities for Black Indigenous People of Color (BIPOC). More work needs to be done to effectively connect the leading guidance on these issues with potential solutions that urban agriculture can offer through systems thinking research strategies and practice.

For example, future research that calculates the potential to sequester atmospheric carbon dioxide into urban farm soil by adopting, analyzing and assessing Best Management Practices (BMPs) for soil carbon storage and using new tools to measure soil health presented in the *Massachusetts Healthy Soil Action Plan* (Zaltzberg-Drezdahl and Newman 2020) would address a critical knowledge gap and make a huge contribution to urban agriculture's list of known, valuable ecological benefits. It is also an excellent opportunity for Participatory Action Research (PAR) in which there is growing interest among practitioners and scientists due to the mutual knowledge and skill building benefits that support ecological community land management and urban farming practices (Gregory and Peters 2018).

Future research is also needed to help identify strategies to shorten local food supply chains in order to make them more resilient to, and quicker to recover from, major disruptions such as those stemming from the recent COVID-19 pandemic, which led to food scarcity, oversupply, and waste, and exacerbated extant food access disparities (Pitcoff et al. 2020). Knowing that these impacts will be worsened by climate change, food system assessments are needed in every community to help local producers design and plan their farms to suit local consumers, to identify and prioritize needed food storage and processing infrastructure, and to help ensure access to fresh, healthy, local food for all residents with the help of food and nutrition benefit programs like SNAP and HIP.[1]

Future practice in teaching would benefit from new curricula geared for high school-aged and elementary school students. There is also a need for innovative engagement strategies to support the learning of college students with varying

[1] https://www.mass.gov/service-details/healthy-incentives-program-hip-for--clients#:~:text=The%20Healthy%20Incentives%20Program%20(HIP,up%20to%20a%20monthly%20limit.

degrees of educational maturity and stamina for scholarship. For instance, strategies that engage students by providing clear structure and direction will be needed by some students, while others will thrive with a supported, self-directed approach to learning. Continuing to offer self-directed and experiential learning approaches in which students investigate topics that align with their career interests and have the opportunity to gain insight on potential careers paths is vital to sustainability education and professional practice. Connecting with real-life practitioners can be accomplished by having students conduct case study research involving interviews and data gathering, through volunteer work, and by inviting practitioners into the class setting as guest lecturers presenting their work.

A foundational learning exercise that teachers could undertake at the outset of a course with students of any age group is to develop new glossaries of current urban agriculture terminology. This practice would directly engage students with the fundamental language and concepts used in urban agriculture, including specialized terms and new keywords that emerge as the profession continues to evolve. The exercise would be a useful way to equip the class with a shared language for the group's learning and would support the task of identifying current value metrics and assessment criteria which students could use in their research later on in the course (Zaltzberg-Drezdahl and Newman 2020).

Although the focus of this chapter is on applying agroecological systems thinking to urban agriculture education and research at the college level, the approach is also meant to be replicated across a range of sustainable farming and food systems projects by actors working individually and in partnership to achieve mutually beneficial goals and to advance the use of new urban agriculture research, knowledge, and action in policy and community development around the world.

References

Burke, L. (2020, October). *As college staff face layoffs, some argue against budget cuts.* Inside Higher Ed. https://www.insidehighered.com

De Schutter, O. (2010). *Report submitted by the Special Rapporteur on the right to food.* United Nations General Assembly. (A/HRC/16/49). http://www.srfood.org/images/stories/pdf/officialreports/20110308_a-hrc-16-49_agroecology_en.pdf

Esters, L. T., & Bowen, B. E. (2005). Factors influencing career choices of urban agriculture education students. *Journal of Agricultural Education, 46*(2), 24–35. https://citeseerx.ist.psu.edu/viewdoc/download?doi=10.1.1.569.6330&rep=rep1&type=pdf.

Farrell, H. K. (2011). *Illuminating urban agriculture: A new framework for understanding complexity* (Master's Thesis, University of Massachusetts). Scholarworks@UMassAmherst. http://scholarworks.umass.edu/larp_ms_projects/29/

Fritsch, J. M. (2013). Urban agriculture programs on the rise. *Techniques: Connecting Education & Careers, 88*(2), 20. https://www.questia.com/magazine/1G1-320734820/urban-agriculture-programs-on-the-rise-agriculture.

Gliessman, S. R. (2020). Transforming food and agriculture systems with agroecology. *Agriculture and Human Values, 37*, 547–548. https://link.springer.com/article/10.1007%2Fs10460-020-10058-0.

Gregory, M. M., & Peters, S. J. (2018). Participatory research for scientific, educational, and community benefits: a case study from Brooklyn community gardens. *Journal of Agriculture, Food Systems, and Community Development, 8*(A), 237–259. https://doi.org/10.5304/jafscd.2018.08A.010.

Hilimire, K., Gillon, S., McLaughlin, B., Dowd-Uribe, B., & Monsen, K. (2014). Food for thought: developing curricula for sustainable food systems education programs. *Agroecology and Sustainable Food Systems, 38*(6), 722–743. https://www.researchgate.net/publication/262583577_Food_for_Thought_Developing_Curricula_for_Sustainable_Food_Systems_Education_Programs.

Jacke, D., & Toensmeier, E. (2005). *Edible forest gardens: Ecological design and practice for temperate climate permaculture*. (Vol.#2). Chelsea Green. http://library.uniteddiversity.coop/Permaculture/Agroforestry/Forest_Gardens/Edible_Forest_Gardens_Vol.2-Design_and_Practice.pdf

Laine, E. (2011). *Agroecology and permaculture in sustainable agricultural development: Permaculture farm system case studies in the United Kingdom*. (Unpublished Master's Thesis, University of Edinburgh)

Leahey, E., Barringer, S. N., & Ring-Ramirez, M. (2019). Universities' structural commitment to interdisciplinary research. *Scientometrics, 118*, 891–919. https://doi.org/10.1007/s11192-018-2992-3.

Martens, K. (2005). Great opportunities in urban agriculture: Planning for teaching in non-traditional programs. *The Agricultural Education Magazine, 77*(5), 10–11.

McClintock, N. (2014). Radical, reformist, and garden-variety neoliberal: Coming to terms with urban agriculture's contradictions. *Local Environment, 19*(2), 147–171.

Mougeot, L. (2000). Urban agriculture: Definition, presence, potentials and risks, and policy challenges. *Cities Feeding People Report, 31*. https://idl-bnc-idrc.dspacedirect.org/bitstream/handle/10625/26429/117785.pdf?sequence=12.

Mujčinović, A., Čadro, S., Uzunović, M., Makaš, M., Glamočlija, P., & Drkenda, P. (2019). Entrepreneurial education skills in urban agriculture of Bosnia and Herzegovina. *International Scientific-Expert Conference on Agriculture and Food Industry, (LXIV). https://www.researchgate.net/publication/338775850_ENTREPRENEURIAL_EDUCATION_SKILLS_IN_URBAN_AGRICULTURE_OF_BOSNIA_AND_HERZEGOVINA*

Nugent, R. (1999). Measuring the sustainability of urban agriculture. In M.K. Editor & R. M. Editor (Eds.), *For hunger-proof cities: Sustainable urban food systems* (pp. 95–99). International Development Research Centre. https://idl-bnc-idrc.dspacedirect.org/bitstream/handle/10625/23475/IDL-23475.pdf?sequence=1

Nugent, R. (2000). *The impact of urban agriculture on the household and local economies*. Growing Cities, Growing Food: Urban Agriculture on the Policy Agenda. https://www.researchgate.net/publication/237237115_The_impact_of_urban_agriculture_on_the_household_and_local_economies

Palacios, C., & Elizabeth, C. (2013). *Urban agriculture in Quito: analysis of the sustainability of the gardens of three projects* (Master's Thesis, Flacso headquarters Ecuador). http://repositorio.flacsoandes.edu.ec/handle/10469/6801

Parr, D. M., Trexler, C. J., Khanna, N. R., & Battisti, B. T. (2007). Designing sustainable agriculture education: Academics' suggestions for an undergraduate curriculum at a land grant university. *Agriculture and Human Values, 24*(4), 523–533. https://link.springer.com/article/10.1007/s10460-007-9084-y.

Pearson, L. J. (2010). Sustainable urban agriculture: Stocktake and opportunities. *International Journal of Agricultural Sustainability, 8*(1/2), 7–19. https://www.tandfonline.com/doi/abs/10.3763/ijas.2009.0468.

Philips, A. (2013). *Designing urban agriculture: A complete guide to the planning, design, construction, maintenance, and management of edible landscapes*. Wiley. https://www.wiley.com/en-us/Designing+Urban+Agriculture%3A+A+Complete+Guide+to+the+Planning%2

C+Design%2C+Construction%2C+Maintenance+and+Management+of+Edible+Landscapes-p-9781118073834

Pitcoff, W., Peats, B., Miller, R., & Cole, J. (2020). *Massachusetts' local food system: perspectives on resilience and recovery.* Massachusetts Food System Collaborative. https://mafoodsystem.org/media/projects/pdfs/MALocalFoodPerspectives.pdf

Rangarajan, A., & Riordan, M. (2019). *The promise of urban agriculture: National study of commercial farming in urban areas.* United States Department of Agriculture/Agricultural Marketing Service and Cornell University Small Farms Program. https://smallfarms.cornell.edu/wp-content/uploads/2019/12/Promise-of-Urban-Ag_Full_102919-1.pdf

Vincent, S., Roberts, J. T. & Mulkey, S. (2015). Interdisciplinary environmental and sustainability education: islands of progress in a sea of dysfunction. *Journal of Environmental Studies and Sciences*. https://www.researchgate.net/publication/281536058

Zaltzberg-Drezdahl, K. & Newman, J. (2020). *The Massachusetts healthy soils action plan.* Massachusetts Executive Office of Energy and Environmental Affairs. https://www.ecolandscaping.org/02/developing-healthy-landscapes/soil/the-massachusetts-healthy-soils-action-plan-overview-survey/#:~:text=In%20the%20Fall%20of%202020,Commonwealth%20through%20exceptional%20soil%20stewardship

Chapter 4
Engaging Nashville's Youth in Farming, Food Choice, and Food Access Issues: Two Programs by a Nashville Nonprofit

Michelle Wooten and Josh Corlew

Abstract Food provides a rich topic around which to garner youths' interest since everyone eats. In working to support schools and after-school service providers in Nashville, Tennessee, the nonprofit organization Hands On Nashville (HON) became aware of frustrations that both students and educators have with the limitations of teaching about healthy eating practices and food issues in the classroom. Building on the belief that experiential learning is an important way that children learn, HON developed programs that used agriculture as a lens to teach about healthy eating practices and food issues. In this chapter, implementation and short-term outcomes for two HON programs with distinct target audiences are discussed. Crop City, a youth development program aimed at teaching healthy eating practices to middle school youth, used games and kinesthetic activities in the garden. The Community and Food Internship was aimed at increasing high school students' confidence in growing their own food as well as understanding the complex problems producing and produced by the US food system. Both of these programs demonstrated positive effects, enabling participants to think more holistically about the ways in which students interact with food, their peers, their larger community, and ultimately the earth.

Keywords Food justice · Participatory evaluation · Youth leadership

Michelle Wooten and Josh Corlew contributed equally.

M. Wooten (✉)
Astrophysical and Planetary Sciences Department, University of Colorado Boulder, Boulder, CO, USA
e-mail: Michelle.Wooten@colorado.edu

J. Corlew
Independent Scholar, Nashville, TN, USA

I. DeCoito et al. (eds.), *Teaching and Learning in Urban Agricultural Community Contexts*, Urban Agriculture,
https://doi.org/10.1007/978-3-030-72888-5_4

4.1 Introduction

The two programs discussed in this chapter, Crop City and the Community Foods Internship (CFI), stemmed from the work of an employee hired by Hands On Nashville (HON) in 2010 in the aftermath of a record breaking flood in Nashville. Many previously residential properties had been zoned as floodplains, making development on those spaces unprofitable at best. Many of these properties were blighted and abandoned, making them prime sites for illegal acts such as dumping, prostitution, drug use, etc. Additionally, the maintenance of these unused spaces fell to the city, requiring tax dollars to upkeep a space that neighbors feared would drive property values down and foster criminal activities.

Through a partnership with the local government, HON obtained land access to convert one of the floodplains into an urban farm to aid in its ecological restoration and become a community asset. In its initial design, the space was envisioned as a production site for increasing the accessibility of organic, local produce to urban residents. Later, HON realized the need and potential for using the space to educate urban youth about the benefits of healthy eating. Recognizing that the United States food system provided inequitable benefits and costs along racial lines (Larson et al. 2013), the program leader (second author) sought to make explicit the connections of the food system to other systems such as public education, economic opportunities, and the prison industrial complex. Through a variety of programs, the urban farm became a place where both plants and minds could thrive by centering on a concept that everyone can understand: food. This chapter discusses two HON programs connected to the urban farm, Crop City and the Community and Food Internship.

In the first part of this chapter, we outline Crop City's origins and goals which used urban agriculture to teach middle school youth about sustainable consumption behaviors. Next, we discuss Crop City's programmatic structure and delivery methods as well as participant outcomes from the summer of 2015. We finish this section by providing an evaluation of the program, namely the lessons learned and suggested areas for future research.

In the second part of this chapter, we examine how the Community and Foods Internship involved using urban agriculture to increase interns' self-reliance through critical thought and specific urban agriculture skills. We explore the program's history, the methods of intervention, and evaluate the short-term outcomes in light of the areas that program participants self-identified as meaningful. We interpret the results of this program in terms of the benefits perceived by program participants, with a focus on how the participatory program evaluation methods used may have supported these short-term outcomes.

4.2 Crop City

4.2.1 Overview

Crop City was an activity-based youth development program that attempted to positively impact young people's knowledge, skills, and behaviors about healthy eating and nutrition. Through an interactive curriculum, youth saw first-hand the process of vegetable production from planting seeds all the way to consumption while learning gardening techniques, nutritional literacy, and cooking skills. The subtext of this programming gradually exposed participants to often overlooked aspects of the United States food system including the creation and perpetuation of food deserts, the working conditions of minority farm workers, and environmental impacts. In addition, Crop City highly valued experiential and exploratory learning styles which are especially significant in the context of a public education system that relies heavily on standardized testing and lecture-based instruction.

To accomplish these educational goals, in summer 2015, Crop City operated on eighteen learning objectives that were separated into three categories: growing (gardening knowledge, skills, and experience), nutrition (nutrition and cooking literacy, skills and experience), and impacts (environmental and social impacts of the food system). Each of these eighteen learning objectives can be taught in a multitude of ways which allows for youth to participate in Crop City for multiple years, continuing to gain a concentrated set of concepts without repeatedly engaging in the same activities.

The activities used in Crop City were designed in a modular fashion which allowed for sites with high turnover rates of participants to continue to benefit. With this modular format, participants who were only able to come to one session could still benefit from the experience, while participants that attended additional sessions could find that their combined experiences led towards a more complete understanding of the concepts taught.

Crop City has been employed in two slightly different formats: school year and summer. Throughout the summer, groups of youth participants attended Crop City every week for up to six unique weeks of programming. During the school year participants sometimes participated every week for up to 24 weeks rather than the 6-week maximum during the summer. Because of this extended exposure, spring and fall Crop City activities covered concepts in greater depth. In this chapter, only the summer version of Crop City is discussed.

One of the most unique features of Crop City is its youth leadership component. HON hired a cohort of high-school youth to run the program as youth leaders. Youth leaders trained for over 40-h, learning the curriculum, teaching skills, and peer leadership techniques while gaining teambuilding experience. During the program, the youth leaders were instrumental in the success and evaluation of the program in a number of ways. They enabled higher enrollment through increased leadership presence. They also recorded the experiences of the participants and offered feedback on each activity's engagement and accessibility, often contributing to changes in the curriculum.

4.2.2 *Participant Population*

The nonprofit HON's main function to meet community need through volunteerism involved working closely with many other nonprofits and service agencies in Nashville. These relationships led to increased awareness and enrollment of youth participants in HON programs, because these organizations also offered free enrichment programming to client bases that matched Crop City's target audiences, which will be described shortly. In particular, these organizations brought between 10 and 60 youth at a time to the urban farm.

Since Crop City's program objectives were to increase participants' knowledge, skills, and preference for using whole foods, its target population included youth experiencing the negative impacts of the heavily processed food system in America. Income and race are currently strong indicators of this population (Kershaw et al. 2019; Quinlan 2013; Walker et al. 2010), so these demographics are used to define Crop City's preferred target population. However, HON's partner organizations (e.g., YMCA, Martha O'Bryan Center) each had their own criteria for determining participant eligibility which were used to make a broad assessment of Crop City's participants demographics.

4.2.3 *Origins*

The first iteration of Crop City began in 2012 just after the launch of HON's urban farm to address its proposed usage for educational programming. One of the initial goals of the program was to introduce urban students to the sources of their food, since the urban environment may not offer opportunities to witness food production processes. After the first year, program leaders considered that curricular themes could provide clarity to the program goals and facilitation. Thus, in 2013, a three-station model was invoked, with youth groups rotating between composting, gardening, and nutrition. Feedback from participants showed that the compost activities were by far the least enjoyable part of the program which informed the next year's major shift. In 2014, an "impact" station replaced the composting portion of the program, with a focus on healthy eating skills. A group of dietetic associates from a local university began providing assistance in improving the nutritional focus and using evidence-based practices.

4.2.4 Learning Objectives

In the summer of 2015, Crop City leaders solidified the intended learning objectives into 18 measurable concepts, one for each station – Growing, Nutrition, and Impact – for each of the six unique weeks of programming. Each of the eighteen learning objectives related to the Growing, Nutrition, and Impact stations had a set of measurable indicators that were used to gauge participants' comprehension. Youth leaders' measurement of outcomes were used during daily debriefs to revise or completely redesign individual activities. We now discuss an example learning objective and its specific indicators for each station.

The overarching Growing learning objective was for youth to gain confidence and skill sets required to grow their own food. An example activity to enable under these objectives included having youth learn to sow seeds, transplant seedlings, and care for a plant during its life cycle. The specific indicators used to evaluate achievement of this objective were a participant's ability to sow seeds according to seed packet instructions, demonstrate transplanting with proper depth and spacing, determine necessity for watering and apply correct watering technique, identify "crops" versus "weeds", and phytopomorphize the life cycle of an annual vegetable plant.

The overarching Nutrition learning objective was for youth to learn a set of tools, techniques, and concepts to make nutritious and delicious food choices, feel confident preparing fresh, unprocessed foods beyond the HON learning space, and choose more fruits and vegetables in their diet. Indicators used to evaluate achievement of this objective included a participant's ability to identify one whole food and one processed food and the health implication of a particular food choice.

The overarching Impact learning objective was for youth to understand the interwoven connections between the food system and environmental, social, and economic injustices and make conscious choices that reflect this awareness. An indicator used to evaluate achievement of this objective was a participant's ability to identify one resource used in agriculture and how it is used in industrial agriculture versus sustainable agriculture.

4.2.5 Curriculum

A typical Crop City experience began with an ice-breaker activity. Sometimes this activity related to the day's learning objectives, but often it was a game that encouraged movement, creativity, and provided all participants an intentional space to give input and be heard. After the icebreaker, facilitators led an activity or game to deliver the day's learning objective. These activities were correlated with at least one learning objective. For example, if the learning objective of the day was "Youth will learn the difference between processed and whole food" in order to choose more real foods in their diet, instructors may have used a game called "Cup Full." This game began with the facilitators asking participants to think of their favorite snack food and then calling on several participants to share. The activity then engaged participants in the processes that remove the nutrition from the snack foods original source through a kinesthetic activity. Finally, the HON team found that conducting consistent starting and ending activities during Crop City modules enabled a sense of program ritual amongst participants. At the end of every lesson a participatory game or reflective discussion was used as a process for participants to come to closure. It was also the time when youth leaders recorded which participants retained understanding of concepts encountered in the lessons.

HON invoked a plethora of learning activities – kinesthetic, dialogic, and written – into their curriculum for two main reasons. Firstly, educational scholars consider learning to be an embodied experience, not merely cognitive, and informed by practices and relationships construed in the learning environment (Sengupta et al. 2019; Varela et al. 2017). Individual learning styles vary greatly, and catering to these differences augment learning and retention (Feinstein 2014). Secondly, the United States education system currently emphasizes standardized testing, and consequently classroom instruction primarily takes the form of reading, writing, listening, and speaking which are lingual forms of instruction and evaluation. The exclusive use of these methods can make learning challenging for participants who benefit from other kinds of learning activities. Changing the kinds of activities in which students participated also supported students' repeated enrollment in the program, as different activities could be used to teach each station's focus (Growing, Nutrition, Impact).

Finally, a major objective for every Crop City experience was for participants to have the opportunity to prepare and taste fresh, whole foods. HON noticed that youth who were involved in the growing process of vegetables through planting, weeding, watering, and/or harvesting were more likely to be willing to try these vegetables. Even when weather or other circumstances prevented young people from actively gardening in a session, bringing the raw ingredients for them to prepare played a similar function. While careful supervision was required for youth using cooking equipment, allowing them to practice basic cooking skills seemed to increase their willingness to eat healthy snacks that would otherwise be deemed "nasty" or "disgusting." Even the simple act of ripping a leaf of kale into bite-sized pieces increased a participant's willingness to try a salad.

4.2.6 Crop City Program Evaluation Results

The effectiveness of Crop City was determined by measuring how well participants were able to acquire and retain the learning objectives. HON used a variety of different tools including written surveys, games, group discussions, and observed activities which mirror many of the techniques used to convey the information initially. In the summer of 2015, participants came from six summer enrichment programs. A total of 939 Crop City experiences were provided, each lasting between 2- to 3- h, with a minimum of 30-min spent on each learning objective.

The learning process was facilitated and tracked by youth leaders working in groups of four to lead up to twenty participants through the activities and measuring activities. Youth leaders marked on a score sheet the number of participants in their small group attending and meeting each objective at each station on each day of the program. Results from each week's learning objectives were compiled to generate the cumulative index of success in each objective visible in Table 4.1.

These data resulting from 2015 Summer Crop City show that on average 90% of participants successfully demonstrated knowledge of the learning objectives.

In terms of expenses, the cost of participation per participant per module was $43.88. This was calculated in terms of supplies ($2482), youth leader stipends ($17,452), transportation ($6869), staffing ($14,400 for 2 staff working 720 h at $20 per hour) needed for 939 participants.

4.2.7 Learning Outcomes

Funding, staffing, and temporal limitations prevented HON from tracking individual data across the summer. Subsequently, only post-module measurements were available for the summer of 2015. While these measurements boast high retention rates directly after module instruction, they do not offer a perspective of change in participants' understanding of concepts that could be offered through pre- and post-module measurement.

While HON considers the 90% success rate as extremely positive, the program leader (second author) is aware of the barriers in translating this knowledge into behavior changes. Programs like Crop City will be most impactful when combined with other interventions that allow the knowledge gained through the learning objectives to be applied by the participants. For example, many participants live in publicly subsidized housing units that do not have space available for individual gardening. Therefore, the Crop City curriculum is believed to be best delivered when participants can also participate in a community gardening program or an extra-curricular garden club at their school.

4.2.8 Cost Analysis

Much of the program expenses were associated with staff – youth leader stipends and program leaders' time – as well as transportation. If a community-based and/or volunteer-centric model were used instead, much of the expenses could be eliminated. The actual supplies required to implement the program are nominal, clocking in at just over 6% of the cost. The program leader hopes to make these activities documented and searchable within the next year and subsequently train community volunteers in the process of creating a syllabus filled with tested lesson plans and who can deliver this content at any location with minimal support.

Table 4.1 Cumulative scores

Module	Growing	Impacts	Nutrition
1	83% (134/162)	83% (134/162)	90% (145/162)
2	94% (166/176)	85% (106/125)	95% (168/176)
3	95% (186/196)	89% (174/196)	83% (162/196)
4	91% (97/107)	88% (94/107)	82% (88/107)
5	93% (151/162)	98% (159/162)	97% (157/162)
6	86% (88/102)	86% (88/102)	74% (75/102)

Note: The numbers in parenthesis indicate the number of participants retaining the module's objectives divided by the number of participants attending the module

4.2.9 Implementation Considerations

One of the most basic lessons HON learned through the development of Crop City was that community involvement from the beginning is important. The first two years of the program were focused on intersecting youth with urban agriculture as a way to shift the locality of food production from exclusively rural towards urban growing as a way to supplement food sources. While the authors still believe this is an important issue to resolve for a sustainable food system, the more immediate need – which is a pre-requisite for increasing the production and consumption of healthy foods – is creating value for nutrient-dense and whole foods in urban culture. For populations experiencing generational or cyclical poverty, HON hypothesized that this value shift begins by re-establishing a connection between food and the earth. Gardening is a way to do both of those things. To this end, Crop City's modularity also made the program adaptable to many other contexts beyond HON's urban farm through implementation of the school year curriculum on the property of partner organizations, including schools.

4.2.10 Future Research

While some evidence exists that Crop City impacts participants' knowledge and skills regarding nutrition and gardening, the effectiveness of Crop City to change diet patterns has yet to be confirmed with any reliability. This reality exists for many reasons. One reason is the difficulty of tracking participant behavior outside of their interactions with HON staff. Evaluation data on what participants eat relies on self-reporting. Even with accurate short-term information like this, a longitudinal study on the lasting impact of this knowledge is needed. Additionally, the age of participants makes it unlikely that they are a primary decision maker about the food that is in their home. Even if a child reports a preference for whole foods to processed foods, the likelihood of that preference impacting behavior is low for this reason. This intervention is not meant to be a stand-alone cure for food insecurity. Rather,

Crop City addresses a necessary educational element required to make a holistic change in the food culture of places that have been historically disenfranchised.

4.3 Communities and Food Internship

4.3.1 Overview

The Communities and Food Internship (CFI) is a high school program whose duration spans the school year and is designed to impart interns with the skills necessary to produce all of one's own food as well as provide a working knowledge of the United States food system. In the rendition of the program described in this section, which took place during the 2014–2015 school year, any high schooler in Nashville was eligible to be a participant, considering their attendance to an interview following their submission of an application to the program, and participation in a week of orientation at the beginning of the school year. Interns met together for one-hour every week at either HON's urban farm or headquarters, the latter of which is located in downtown Nashville.

4.3.2 Origins

The CFI originated from the program leader's (second author) desire to expand the leadership opportunities provided to high school students by HON. Inspired by the Academic Year Program of the Boston Food Project, the CFI was intended for high school students that were interested in making positive social change through the food system by increasing their awareness of social inequities created by the United States food system. This occurred through sparking discussion over possible avenues to address these inequities and eventually designing project proposals to address the concerns stemming from these discussions. Initially the program was intended as a professional development program for the purpose of training interns to facilitate Crop City at HON's urban farm (the program discussed in the previous section of this chapter). Ultimately, however, a standalone internship was developed in which participants could choose whether or not to continue as Crop City leaders. The program leader envisioned that after completing the CFI, interns would have a firm grounding in both agricultural skills and the social and environmental implications of the United States food system, important to serve as Crop City leaders.

4.3.3 Curriculum

To address the first goal of maintaining one's own garden, during the fall and spring while weather permits, interns learned garden planning and implementation skills at the HON urban farm. These skills included year-round growing, sustainable fertilizers, food preservation, fruit tree pruning, and annual and perennial planting and harvest schedules, to name a few (Table 4.2). These specific activities were chosen by the program leader based upon his own experiences using agricultural skills in sustenance maintenance.

HON's urban farm was a valuable resource in implementing the program. The program leader was able to demonstrate the techniques while explaining the process and the reasoning behind the process. After watching the instructor, the interns mimicked the activity alongside the instructor, asking questions as they arose and practiced for 20–30 minutes or until the task was completed. For example, when learning about thinning plants, the instructor explained that each plant requires a minimum-level of resources such as soil nutrients, water, and solar energy. When crop plants are too close, the resources are divided in a way that prevents any plant from successfully reaching maturity. Next, the instructor demonstrated the spacing required between a type of crop and how to select which plants to remove or to leave. Interns then worked alongside the instructor, asking questions until they felt confident to perform the task without supervision. Finally, the interns would finish the task and the instructor would review the work to ensure accuracy.

To address the second goal of the program, teaching the effects of industrial agriculture on our social and physical communities in inequitable ways, the political and economic factors leading to the production of food deserts and the health problems associated with them were explored through reading, watching documentaries, and discussion. For example, in one of these meetings, a portion of the film *Food Inc.*, discussing the impact of GMO seeds on the food system, was viewed. Afterwards interns participated in a roundtable discussion guided by questions such as, "What are the implications for patenting living things?" "How do we compare the increase of production to the damage caused by GMOs?" "How can we quantify environmental and social damage?" In a following session, interns were asked to find articles on the benefits of using GMOs in agriculture. During the next meeting these articles were discussed in light of the understanding gained from the previous GMO discussion.

Examples of positive alternatives to industrial agricultural practices were explored to inspire interns to think outside of conventional agricultural practices and pre-existing logistic systems including but not limited to processing, packaging, transportation, and retail. One example of these solutions was explored by watching the documentary *Greenhorns*, which comprises a series of vignettes that showcase the innovative, creative, and inspiring work of many young farmers across America. All together, these discussion-oriented sessions took a similar format to the sessions highlighting problems, engaging the interns in lively discussion and brainstorming sessions after watching a section of the film. By guiding interns through arguments

Table 4.2 Schedule of Activities

Week	Month	Lesson Title
1	August	"A Place at the Table" (fsu)
2	August	Fall planting (ag)
3	August	Soil testing (ag)
4	August	Summer harvesting and preserving (ag)
5	September	Value added fall products (ag)
6	September	Volunteer leading training (fsu)
7	September	Pear picking & preserving (ag)
8	September	Winter cover crops (ag)
9	October	Extending the growing season (ag)
10	October	"Stuffed & Starved"/GMOs (fsu)
11	October	GMOs (fsu)
12	October	Community Bonfire (fsu)
13	November	Food Politics – "Food Inc." (fsu)
14	November	Biodiversity (fsu)
15	November	Fermenting krauts (ag)
16	December	Opposition research techniques (fsu)
17	December	Opposition research debrief (fsu)
18	December	Mid-term evaluation design (fsu)
19	January	Semester recap/Intro. to Solutions (fsu)
20	January	Brainwriting/Defining the Problem (fsu)
21	January	Fair Food Movement (fsu)
22	January	Debating the Food System (fsu)
23	February	Debate debrief (fsu)
24	February	Fruit tree pruning (ag)
25	February	Pea planting, cover crop incorporation (ag)
26	February	Building trellises (ag)
27	March	Transplant greens (ag)
28	March	Potato planting (ag)
29	March	Direct seeding for summer crops (ag)
30	April	Summer cover crops (ag)
31	April	Volunteer leading training (fsu)

(continued)

Table 4.2 (continued)

Week	Month	Lesson Title
32	April	Spring transplanting (ag)
33	April	Water management (ag)
34	May	"Greenhorns" (fsu)
35	May	Intervention Design 1 (fsu)
36	May	Intervention Design 2 (fsu)
37	May	Goodbye party/project presentation (fsu)

Note: (fsu) stands for food system understanding foci while (ag) stands for agricultural skills learning

embedded in the multiple sides of each issue, the program leader hoped to foster interns' nuanced understanding and articulation of barriers to equitable food access, and a drive to action.

Volunteering in high school is an indicator for civic engagement in adulthood (Hart et al. 2007). Consequently, the final part of the program was aimed at developing interns' confidence to share their knowledge and skills, through their design and execution of volunteer-based projects that addressed injustice in the food system. To provide inspiration and support for the creation of these projects, interns learned HON's process of project development and community organization skills.

4.3.4 Evaluation Design

To conduct a program evaluation of the CFI, a doctoral student of research methodologies at The University of Alabama (first author) volunteered her efforts as part of her enrollment in a program evaluation course. To begin collaborating with HON, she interviewed the program leader (second author) and developed a logic model, a common program evaluation practice, which was useful to ascertain which of the outcomes the evaluation could seek to address given the timeframe of the study in spring 2015 (Fig. 4.1).

The program evaluator and program leader agreed upon a qualitative inquiry of the program's short-term outcomes, namely:

1. To invoke the appropriate agriculture knowledge and skills to maintain a garden for a year.
2. To understand the main symptoms and drivers of a broken food system.

Because the internship was discussion and activity-based, they adopted an evaluation approach that corresponded with this kind of learning. In particular, they adopted a practical participatory (Cousins & Earl,1992, as cited in Fitzpatrick et al. 2010) and naturalistic (Guba and Lincoln 1989; Lincoln and Guba 1985) evaluation design. In practical participatory evaluation designs, continuous communication, contact, and collaboration are of prime importance between evaluators and invested stakeholders. Subsequently, the program evaluator made all evaluation-design decisions in collaboration with the program leader. In naturalistic evaluation, evaluators study the program activity as it is intended to occur, with little to no evaluation procedures invoked. Accordingly, evaluators learn about program participants' perspectives on program participation through evaluators' observation and or participation in regular program activities.

To learn about participants' perspectives, the evaluator began to develop familiarity with program participants through occasional visits to CFI meetings during Fall 2014, while interns were learning agricultural skills at the HON urban farm. When the weather cooled, the internship activities moved indoors and became more discussion-oriented in nature. At this point, the program evaluator hosted an interviewing workshop for the interns, in which interns discussed and practiced qualities of a good listener, including asking open-ended questions and follow-up questions. The program leader felt these would be useful skills for interns, especially as they could aid in communicating with persons who had different stances to food issues than they did throughout the remainder of the internship.

Later in the school year, the program evaluator met with small groups of interns to conduct focus group interviews, which served as the primary means of evaluation data sources. Using a participatory approach, interns were asked to spend a few minutes writing questions they would ask fellow interns about the meaning and value of their CFI participation. Interns then met in pairs to vet their developed questions to each other and converse about their responses without any involvement of the program leader or program evaluator. Next the program evaluator asked the interns to repeat their questions and responses, this time sharing their responses with her so that she could record them and ask any follow-up questions, especially as they related to the short-term program evaluation outcomes. These responses were later coded and thematized by the program evaluator, at the request of the program leader, using thematic analysis (Roulston 2010).

4.3.5 *Outcomes*

This section is devoted to describing themes the evaluator interpreted from two pairs of interns' questions and responses to each other. Each pair came up with a different set of questions. Example questions the pairs generated included:

- How do you think we can influence kids with what we've learned this year?
- How do you think differently now about what foods to eat?
- How did you like the canning and picking experience?
- In what ways do you take what you learned and use it in your own life?
- What has been the most meaningful experience so far?
- Has [any other intern] had a similar experience?
- Would you recommend this program to another person?

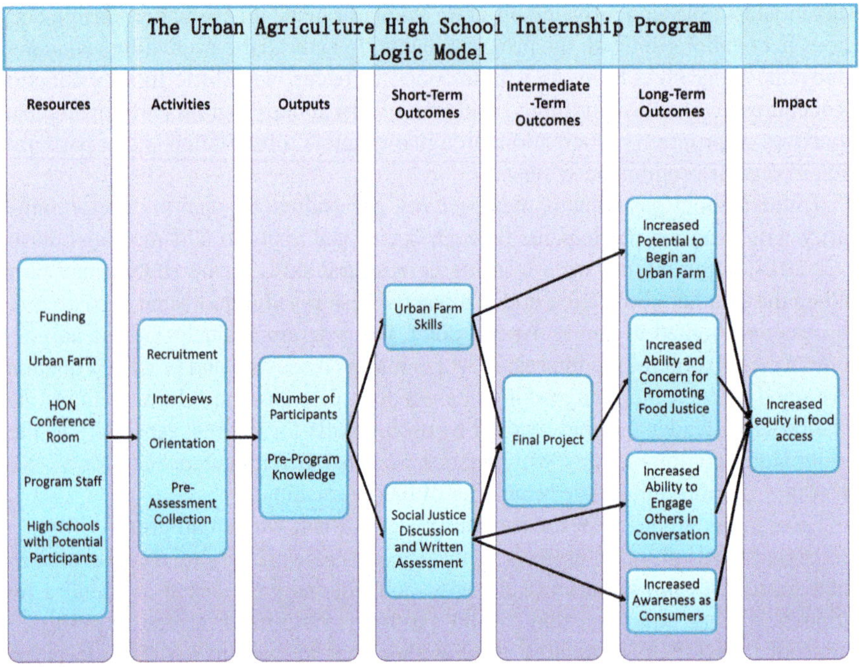

Fig. 4.1 Community and Food Internship Logic Model

Some of the content broached in these questions demonstrated that interns held shared memories, motivations, images, and cognitions that were outside of the program evaluator's purview. For example, the program evaluator was not aware of program activities such as the "canning and picking experience." Nor was she aware that "influencing kids with what we've learned" would be important to the interns. Intern-led inquiry consequently enabled learning about short-term outcomes of CFI participation that might not have been found through evaluator-led inquiry. Themes and anonymized quotes interpreted from interns' focus group interviews are described below. Notably, more personally relevant short-term outcomes were perceived by CFI participants than the two anticipated by the program leader, namely "Agricultural Skills" and "Food Insecurity."

Food Choice The adoption of new thinking and behaviors related to food choice was one of the foremost mentioned topics during interns' conversations. Daily thinking about food healthfulness and costs associated with their food choice were mentioned. Here Claudia and Rosetta describe their valuing of healthy foods in daily life:

Rosetta: We now think about what kinds of things go into your food because that's what we're eating – we're taking it into our bodies. We're more educated now about what we're actually eating – so making the right kind of choices.

Program Evaluator: You said you did that the other day?

Claudia: [laughs] Yeah, last night my family wanted to go to [fast food chain restaurant] – they wanted the fried chicken like Southern food. I asked if we could go to [sit-down chain restaurant] instead because I know what's actually going into my food. It was a healthier choice – than [fast food chain restaurant]. We don't know what they put in the oil or what they put the chicken in.

Rosetta: Like going to the store I actually think about what stuff we're buying and I'm excited because the Farmer's Market is coming back.

Claudia: Yeah, I'm excited!

A habitual concern over food choice was also apparent in Marny and Jacob's narrative, which was more political in nature:

Marny: You just have to be open-minded with that process of "Every dollar I spend is basically me casting a vote" for how I think it should be.

Program Evaluator: So you think about that when you buy stuff?

Marny: Yeah, all the time. That's pretty much all I think about now.

Jacob: It does make you really concerned about what you're buying.

Agricultural Skills One of the most widely cited and valued internship activities by the interns was the development of new agricultural skills. A range of interns' activities were associated with these skills, including learning to compost, plant, rotate crops, cover crops during the winter, and using the harvest to make meals:

Program Evaluator: So you're both going to have your own gardens this year?

Claudia: Well, we've both had gardens, it's just – I would grow the same – I knew what would grow in my yard. But I learned that growing the same thing every summer isn't good for your soil, so I created my own compost and put that down for the winter, and now I'm going to pick out different stuff. So it has influenced me in a positive way in my own house.

Rosetta: Yeah, that's cool! We are too – we're doing tomatoes and I think, onions.

One intern described her perceived low odds for her ability to purchase healthy foods later in life, during which her newfound agricultural skills would become important:

Marny: One of the purposes to me is the agricultural side, because I'm really interested in gardening and that sort of thing. At home I've started building raised beds for my backyard because there's not enough sun in specific parts of my backyard so I've learned a lot through that. I think it's very helpful to have that experience because – probably I'm going to live in a condo – I'm not going to have access to affordable food someday so I'm going to want to be able to grow those things. But that's also helpful because that's something you take with you forever. I mean knowledge in general is something you can't get rid of.... It's nice to be prepared.

Internship Learning Interns repeatedly noted their enjoyment of the practical, hands-on nature of internship learning. Even though it was an extra hour out of their week, often when they were constrained with school work or other extra-curricular activities, they expressed that the aim of the internship, as well as the relatedness experienced with other participants, made every meeting valuable to them. Here, Jacob contrasts school learning with his internship learning that illustrates the value of the internship learning environment in contrast to his school learning environment:

Jacob: Aside from the agricultural skills, I think a skill I've developed is in the way of learning about food policy and working in a group. At school, group work can be difficult because the stakes are really high and very immediate. And I think that here, where the stakes are high and maybe even higher in terms of "we really want to do something meaningful for society" – and it does have a direct effect on real peoples' lives other than yourself.

I think that the stakes aren't as immediate here. At my school, people care a lot about grades. I wouldn't necessarily say that it's a competitive culture, but it is a very college-aware culture, and getting into college is competitive, so even though our administrators try very hard to make it entirely about the learning – and I think it is mostly about the learning – a lot of the time it can get a little bit derailed when you look at the seniors, look at the juniors – I'm a junior now – you know, they've got all these milestones. At my school as a sophomore you start all the AP exams, and you have to get high scores on those in order to get college credit or to get into good colleges or to get money at colleges later on. So it's different here, because here what we're learning has the chance to make a meaningful impact on people but I think that if we messed up one time it doesn't have the potential to ruin everything.

Later Jacob adds that because of the immediate, high stakes learning environment experienced at school, group learning inevitably ends:

Jacob: "Okay, let's just move on." Whereas here, you really want to hear what others have to say. It also doesn't end – whereas at school, it definitely has to end.

Marny subsequently describes how she relates to this sentiment, because at her public school, which has a vocational training focus, grades are given based on rankings:

Marny: We'll have to build something and whose [sic] ever is the most successful gets the highest grade and it just goes down from there. So there's a very small margin for error. And that's given me the tendency to just take over everything in a project because I don't trust other people with my success, which I think is part of what Jacob is trying to say.

Peer Collaboration Interns expressed increased valuing and respecting of peers' opinions and abilities because of the socially meaningful aim of the internship as well as its discussion-based curriculum. The types of interactions experienced in the internship fed into their peer relations at school:

Marny: I've also noticed since I've started this [internship] that at school everyone has different assets and you can work to everyone's strength. Yes, this kid at my school sleeps in class all day but he likes to work with his hands, so I can tell him to build things while I do the logistics of everything.

Jacob, agreeing with Marny over the diversity of internship participants, expressed:

Jacob: It definitely helps to work in a group that is diverse and where somebody else's opinion – that I might not have considered before – always is worthwhile use of time. Even if we don't end up using it, it's a heuristic experience.

Youth Engagement The engagement the interns experienced in these activities ignited interest in imparting their experiences to youth they work with in similar ways:

Claudia: How do you think we can influence kids with what we've learned this year? We can make it helpful for the kids to play games that are hands-on and to learn what we want to teach them in an effective but fun way.
Rosetta: Yeah, I think it's more effective getting the ideas across that we've talked about if we make it fun for them instead of giving them a worksheet.
Claudia: Right, [sarcastically] "read this!"
Rosetta: It's more fun when they actually get to make food and see the food and eat the food. I would like that.
Claudia: Free food is always good for kids. Healthy free food!

All of the interns expressed a passion for engaging youth in the practice of sustainable agriculture and social justice issues related to food access.

Social Skills A number of interns described improved social skills including leadership, competence in public speaking, listening, and maintaining a calm nature when sharing counterpoints to peers' and adults' opinions. One intern shared a personally profound social transformation that she attributed to the internship:

Marny: I used to be really soft-spoken. Actually, I didn't really talk at all. But through this I've been very outspoken and opinionated which has reflected in a lot of aspects in my life. Everyone is like, "Marny, when did you start talking so much?"
Jacob: [laughs]
Marny: "Just now," actually.

Food Insecurity Finally, while all the interns expressed concern with the relationship between personal food purchasing habits and food access injustices, one intern in particular, Jacob, a second-year Community and Foods Intern, described that his most meaningful gains associated with the internship were those related to understanding the presence of food insecurity in the United States. In terms of food purchases, he stated:

Jacob: One problem is that it's really hard to find things that you can buy without feeling unethical because a lot of organic food is flown on planes. It's also really expensive. So there's that motivating factor, too. And food from farmer's markets and small farms – it goes against the theory of comparative advantage that American economics is predicated on so there are a lot of economic dilemmas involved as well – and the moral dilemmas – and a lot of organic food is actually contaminated because the USDA has really relaxed regulations. It's almost like if you're voting with your dollars you're always trying to find the lesser of 5 or 6 evils.

In terms of how the internship influenced his perspective on food access injustice, he noted:

Jacob: I also didn't expect how vast the problem was. Like a lot of the time we'll be having discussions and we'll get to the point where we're like wait – so we're discussing the food system but now we're discussing the revolving door in the US government, so don't we need to fix that too? Some meetings at the end it feels like all of society needs a little bit of a shift. Because I think we realize that things are so interconnected that it's really hard to separate presidential appointees to how the food system looks in inner city Nashville. Even though they might not seem connected.
I think you also have to see how [food insecurity] affects you even if it doesn't affect you immediately, because for a lot of people in Nashville, living in their "bubbles" – you see the affect, "Well, there's not a food desert so it doesn't really affect me." But it does,

> because our schools are definitely big enough to encompass the bubbles, the workplace is big enough to encompass the multiple bubbles. Even if you don't ever see someone in your life, you still want them to have a meaningful and happy existence. I doubt very many people would dispute that statement but it's sometimes hard to see that the effects of what you do in your bubble can detract someone's success in another area of Nashville.

Jacob and Marny both described that the vastness and interconnectedness of the social constructs inducing food insecurities were overwhelming, but that the issues raised in the internship were motivating:

> Program Evaluator: Can you tell me about the overwhelming sense? Is there hopelessness or sadness or just, "Oh my gosh, I never knew all this?"
>
> Marny: I wouldn't say there's negative connotations, I would say it's more motivating than anything for me. Just because of the scope of the work that needs to be done – it's more a sense of urgency of "I have to do this" because it's affecting everyone. So it's pretty serious.
>
> Jacob: I think it's a motivator too. I think a lot of that has to do with how [the program leader] presents it to us. He never just presents a problem. He brings in some potential solutions to get us thinking. It doesn't feel so hopeless because even though there are problems, there are a lot of people who care about them and there is change. I'd like to think that we're on an onward march toward economic equality. Although everybody thought that global trade liberalization would bring that and it didn't, so maybe something else will bring it in our generation or five generations down the line. It doesn't feel helpless. I think that what we're doing has the potential to make it better for those who come at the same time or after us.

4.3.6 Cost Analysis

Outside of staff time, the program costs were negligible, utilizing tools and supplies that were necessary for the maintenance of HON's urban farm regardless of the implementation of the CFI. Specifically, the small amount of printing, the use of writing utensils, and meeting space were all assumed by HON through general operating expenses and were not tracked. Further, the program leader only required approximately 2 h each week outside of the weekly hour-long internship meetings preparing for the program.

4.3.7 Conclusion

The results of the focus group interviews reveal that one-hour weekly meetings over the course of a school year exploring agricultural skills and social justice issues related to food access can have multiple influences on youth understanding of and motivation to address these issues in their own communities. By engaging interns in "fun," relational activities focused on sustainable agriculture to healthy meal creations, interns felt empowered to not only incorporate new practices into their daily living, but to share their new skills with other youth. Further, they began to think continuously about how their own food purchases related to wider food access problems, and how food insecurity was interconnected to other social patterns. These short-term outcomes were much broader than the program leader's anticipated short-term outcomes of being able to maintain a garden for a year and understand socio-political complexities undergirding an unjust food system. Subsequently, such results suggest that clubs or internships patterned on HON's Community and Foods Internship may be instrumental in helping youth connect STEM and social science school concepts learned in school to a practical, community-based outlet.

In addition to noting the powerful and meaningful impact of the internship on its participants, from a methodological standpoint, the participant-led nature of the interviews served as a powerful indicator of the participants' ability – even those of youth – to evaluate their program. Given the freedom to develop their own inquiry into their perceived valuing of the program, participants described nuanced perspectives that may have otherwise been obscured by internal or external leaders' and evaluators' questioning. For example, the program evaluator consistently found herself eliciting further details of what Argyris and Schoen (1978) call "shared images and [organizational] mental models" (as cited in Fitzpatrick et al. 2010) described by interns during their question-and-response dialogues (as cited in Fitzpatrick et al. 2010, p. 205). Further, from a theoretical standpoint, because interns were given the opportunity to question and narrate their own values and experiences, they attained "a pathway into another's life experience and a means of making sense of [their] own" (Yardley 2006). The questions they posed to their interview partner were a form of re-working a sense of self as they collected and added their partner's experiences to their numerous ways of thinking about the program. Therefore, allowing interns to make the evaluation procedure personal was a means to accessing the elements of the program that were important to them and to each other as participants, rather than those elements that were only known or valued by those in charge of the internship and its evaluation.

Acknowledgements We are grateful to Dr. Aaron Kuntz (Florida International University) and Dr. Stacy Hughey-Surman (University of Alabama) for serving in advisory roles toward the program evaluation. We are thankful to the reviewers for their careful and detailed critique which aided in the improvement of this chapter text.

References

Feinstein, S. (2014). *From the brain to the classroom: The encyclopedia of learning*. Greenwood. Retrieved on 12 March 2016 from http://www.myilibrary.com?ID=604705

Fitzpatrick, J. L., Sanders, J. R., & Worthen, B. R. (2010). *Program evaluation: Alternative approaches and practical guidelines* (4th ed.). New York: Pearson Education.

Guba, E. G., & Lincoln, Y. S. (1989). *Fourth generation evaluation*. Thousand Oaks: Sage Publications.

Hart, D., Donnelly, T. M., Youniss, J., & Atkins, R. (2007). High school community service as a predictor of adult voting and volunteering. *American Educational Research Journal, 44*(1), 197–219.

Kershaw, K. N., Klikuszowian, E., Schrader, L., Siddique, J., Van Horn, L., Womack, V. Y., & Zenk, S. N. (2019). Assessment of the influence of food attributes on meal choice selection by socioeconomic status and race/ethnicity among women living in Chicago, USA: A discrete choice experiment. *Appetite, 139*(April), 19–25. https://doi.org/10.1016/j.appet.2019.04.003.

Larson, C., Haushalter, A., Buck, T., Campbell, D., Henderson, T., & Schlundt, D. (2013). Development of a community-sensitive strategy to increase availability of fresh fruits and vegetables in Nashville's urban food deserts, 2010–2012. *Preventing Chronic Disease, 10*, E125. https://doi.org/10.5888/pcd10.130008.

Lincoln, Y. S., & Guba, E. G. (1985). *Naturalistic inquiry*. Thousand Oaks: Sage Publications.

Quinlan, J. J. (2013). Foodborne illness incidence rates and food safety risks for populations of low socioeconomic status and minority race/ethnicity: A review of the literature. *International Journal of Environmental Research and Public Health, 10*(8), 3634–3652. https://doi.org/10.3390/ijerph10083634.

Roulston, K. (2010). *Reflective interviewing: A guide to theory and practice*. Thousand Oaks: Sage.

Sengupta, P., Shanahan, M.-C., & Kim, B. (Eds.). (2019). *Critical, transdisciplinary and embodied approaches in STEM education*. https://doi.org/10.1007/978-3-030-29489-2

Varela, F. J., Thompson, E., & Rosch, E. (2017). *The embodied mind, revised edition: Cognitive science and human experience*. MIT Press.

Walker, R. E., Keane, C. R., & Burke, J. G. (2010). Disparities and access to healthy food in the United States: A review of food deserts literature. *Health and Place*. https://doi.org/10.1016/j.healthplace.2010.04.013

Yardley, A. (2006). Living stories: The role of the researcher in the narration of life. *Forum Qualitative Sozialforschung / Forum: Qualitative Social Research*, *9*(3), [34 paragraphs].

Chapter 5
Resurfacing Environmental Identity in Coastal Peru

Daniela Benavides Reiss, Adriana Gonzalez-Pestana, and Joaquín Leguía

Abstract A Forest for Ancon is an urban reforestation project aimed at creating an environmental tool for children. The goal is to foster environmental literacy and identity in children and to incentivize teachers to deliver outdoor lessons. Human-centered design was used. The forest is located in Peru's capital, Lima, inside a boarding school that houses 400 children. The project planted 450 trees from five different species that are endemic to and play an important ecological role in the Peruvian coastal desert. A diverse group of stakeholders came together to establish the forest: a non-profit ideated the project and executed it in collaboration with the non-profit responsible for housing and educating the children involved in the project, and a corporate bank that provided financing and labor through their staff volunteer program. A revenue stream was created by providing educational tours for other students. Teachers' perspectives and barriers to implementing outdoor lessons have been identified. Moreover, the potential to use this area as therapy for disengaged students is explored. Recommendations for future designs in urban agriculture education are presented: including a multidisciplinary team, creating diverse new learning possibilities, how to make the project financially viable, and how to address barriers according to teachers' perspectives. Finally, we propose how this project can be scaled up.

D. B. Reiss (✉)
ConCiencia NGO, Lima, Peru

NatureForAll, International Union for Conservation of Nature, Gland, Switzerland
e-mail: daniela@conciencia.org.pe

A. Gonzalez-Pestana
ConCiencia NGO, Lima, Peru
e-mail: adriana@conciencia.org.pe

J. Leguía
Asociación para la Niñez y su Ambiente (ANIA), Magdalena del Mar, Lima, Peru
e-mail: jleguia@gmail.com; https://en.aniaorg.pe/

I. DeCoito et al. (eds.), *Teaching and Learning in Urban Agricultural Community Contexts*, Urban Agriculture,
https://doi.org/10.1007/978-3-030-72888-5_5

Keywords Urban agriculture education · Environmental identity · Children · Teachers · Peru · Coastal desert · Endemism · Environmental education · Reforestation · Huarango

5.1 Introduction: Outdoor Education in Peru

And perhaps the link between knowledge and love is magic.
~ David Sobel.

An outpost of magic and greenery rests in the middle of an otherwise barren lot surrounded by sandy desert. Four hundred young trees and 400 children from the boarding home next door connect with each other. The girls choose to call themselves their *fairy godmothers;* the boys, *forest rangers.* The endangered native trees that could vanish soon and these children – who so fully enjoy exploring them and learning about nature through their accessible tools – need each other. And when they meet, those of us who have the privilege of joining them, witness the special magic of seeing nature through children's eyes again.

This chapter is about an outdoor education project established in Peru with the purpose of increasing the green space of a school, modeling lessons that can be taken outdoors and through these, contributing to the students' environmental literacy and identity, among other possible positive results. The hope of this project is that the combination of access to green space and the modeling of outdoor lessons will bring many benefits to the students and teachers (Fig. 5.1). Let us explore why this project is important.

5.2 Urban Agriculture Education

Over half of the world's population lives in urban areas. The UN predicts that 68% of the world's population will be urban by 2050 (UN World Urbanization Prospects 2017). For many urban dwellers, opportunities for engagement with nature and for understanding the foundations of agriculture may be scarce. Urban children become increasingly disconnected from the cycles and processes that sustain life on our planet, providing food and other primary resources.

In urban forestry and orchard projects, visitors and students can observe and experience natural processes and interactions, such as those between the weather, soil, plants, insects, and pathogens. Even basic observations, such as plant growth, can set the foundation to understanding the common traits shared by all living beings. This way, urban agriculture has been proven to be an effective way to reconnect people to nature and is increasingly used as a teaching tool in schools (Kuo 2001). Multiple benefits have been attributed to urban agriculture education. These include improved academic performance and attitudes toward school, increased

Fig. 5.1 Panoramic view of the project site three months after planting. The arid hills typical of the central Peruvian coast can be seen in the background

agricultural literacy and food knowledge, improved health and nutrition outcomes, environmental awareness, emotional development, prosocial behaviors, leadership and skills development (Blair 2009; Hirschi 2015; Robinson 2016; Rogers 2018; Williams and Dixon 2013).

5.3 Importance of Access to Nature

Educators, parents and researchers worry about the alienation of children from nature processes and their life-support systems (Louv 2006). The last report from the Children and Nature Network highlights several research studies that correlate time spent in nature with better health, well-being, and development of core skills, such as problem-solving and creativity, community engagement and happiness in children (Charles et al. 2018). Also, a comprehensive survey conducted with teachers in 45 countries across the world concluded that children are more engaged in learning (88%), better able to concentrate (68%), and better behaved (65%) when lessons are taken outdoors (Project Dirt 2018).

5.4 Environmental Identity

Environmental identity, also known as ecological identity or ecological self, is defined as the perception of oneself as part of the natural environment. It is a sense of connection to some part of the nonhuman natural environment, based on history, emotional attachment, and/or similarity, that affects the ways in which we perceive and act toward the world; a belief that the environment is important to us and an important part of who we are (Clayton 2013).

Understanding environmental identity is thought to be pivotal to discovering how to successfully cultivate pro-environmental values that support the environment (Payne 2001). Childhood experiences in nature that are positive, regular and meaningful are the strongest predictors for pro-environmental attitudes and lead to children pursuing environmental professions in adulthood (Charles et al. 2018; Chawla 1998; Chawla and Cushing 2007;). Evidence suggests that nature-rich cities and urban forest areas are an essential part of a long-term strategy to care for the Earth (Guiney and Oberhauser 2009). Furthermore, positive environmental identity contributes to well-being (Olivos and Clayton 2017; Wolsko and Lindberg 2013).

Environmental identity is facilitated by experiences in nature (Chawla and Derr 2012; Tam 2013), and develops through interactions with the social environment (Wortham 2006). Experiences of nature in the company of others affect our understanding of what nature signifies as well as the way we conceptualize our own relationship with nature (Olivos and Clayton 2017).

Children generally lose interest in school science as they move toward middle school (Osborne et al. 2003; Thomson and Fleming 2004) and school science programs fail to embrace students' interests in and feelings about the natural world (Calabrese-Barton and Yang 2000; Carrier et al. 2013). Since students attach spiritual or aesthetic meaning to the natural world (Cobern 2000), environmental identity can increase children's interest in school science (Blatt 2013; Tugurian and Carrier 2017).

5.5 Environmental Literacy

Environmental literacy is a measure of a person's understanding of the environment – at a local level or global scale. Both anecdotal and investigative evidence indicate that educational approaches that build environmental literacy positively affect environmental attitudes and reported behavior (Ardoin 2013). Research also correlate education approaches that build environmental literacy to long-term impact (Volk and Cheak 2003). Follow-up studies of one such project in Molokai Island, Hawaii revealed that the students became involved in community decision making involving green spaces (Volk and Cheak 2003). Evidence suggests that nature-rich cities and urban forest areas are an essential part of a long-term strategy to care for the Earth (Charles et al. 2018; Guiney and Oberhauser 2009).

5.6 The Project: A Forest for Ancon

A Forest for Ancon is an urban reforestation project created with the ultimate goal of developing an environmental education tool. The project was the result of several local actors working together: A non-profit ideated the project and executed the design, in benefit of another non-profit in charge of the housing and education of a few hundred children. A corporate bank was involved in providing financing and labor through their staff volunteer day. The partners planted around 450 juvenile trees of five tree species endemic to the desertic Peruvian coast, specifically their local names are "Algarrobo" *(Prosopis pallida),* "Huarango" *(Prosopis limensis)* "Molle" (*Schinus molle*), "Huaranhuay" (*Tecoma sambucifolia*), and "Tara" (*Caesalpinia spinosa*).

As a result, the children are taught about these endemic species and encouraged to identify them as a part of the local history. The area provides a small haven for the many local migratory birds by restoring some of the endemic vegetation in an area with scarce greenery. As other living beings start to appear in the resulting ecosystem, the teaching opportunities expand into entomology and zoology. Another objective was to understand how to use the forest as a case study for nature-based education. This is an ongoing objective in which the forest represents an important element for teachers to engage children in science-based learning activities. Aside from the original goal, foresting this area protects the unused lot from dumping and land invasion.

5.7 Background on the Location

The Forest for Ancon project is uniquely situated in the Sechura desert, an extremely arid zone with 100–200 mm of precipitation per year (World Wildlife Fund 2014). Tree species have adapted to this harsh ecosystem by becoming highly specialized. The most iconic of these are the "Algarrobo" (*Prosopis pallida)* and "Huarango" *(Prosopis limensis).* They can live for over a millennium and have the deepest known root systems that can reach 70 m long (Díaz 1995; Stone and Kalisz 1991). These are keystone species that enhance soil moisture, fertility, and desalination, while also integrating fragile desert ecosystems and countering erosion, thus moderating desert environmental extremes and enhancing biodiversity (Beresford-Jones et al. 2009a). Their survival depends on the occurrence of El Niño oceanographic and atmospheric events and the strong rainfalls and floods that it brings. This way, El Niño regenerates the dry forest in a non-catastrophic way. Additionally, it brings an abundance of nutrients to the soil, replenishes groundwater levels, and deposits alluvial sediment over extensive areas, thus providing conditions for floodplain agriculture. In this environmental condition, *Prosopis* plays a crucial role by maintaining potentially erosive flow within river channels and by preserving irrigation systems (Beresford-Jones et al. 2009a).

The dry forest in coastal Peru sustained ancient human settlements by providing food, firewood and forage. Perhaps as much as 50% of the human diet in the area was composed of wild plants, including fruit of the Huarango (Beresford-Jones et al. 2009b; Beresford-Jones 2011). Nonetheless, the Nazca culture in southern Peru (500 A.D.) deforested the Huarango forest, triggering its own demise. An El Niño event brought the collapse of this ancient civilization, causing destructive flooding, erosion, and desertification without the countering effect of Huarango (Beresford-Jones et al. 2009a, b; Beresford-Jones 2011).

The Peruvian population was concentrated in the Andes until the mid to late twentieth century. In 1940, 65% of the Peruvian population was located in the Andes; in 2007, this had dropped to 32%, while 54.6% is now located on the coast (International Organization for Migration 2015).

Lima's population soared from 828,298 in 1940, to nearly ten million in 2015, which represents 28.4% of the total Peruvian population (INEI 2015). This constant and massive migration to the coast has jeopardized the dry forest, not only due to clearing for urbanization but also due to wood extraction. The Huarango has become used almost to extinction as charcoal to roast chicken with a signature smoky taste, and to provide fuel for a burgeoning Pisco distilling industry. As a result, the Huarango is endangered with only a small fraction of the original Huarango forest habitat remaining.

In Ancon, where the project is situated, the riparian dry forest and coastal lomas (seasonal fog meadows in coastal desert) has been urbanized (Fig. 5.2). The location is situated on the path of the heavy northern-bound transit exiting the capital city. Scarce green areas remain – Lima presents 3.7 m^2 of green areas per habitant, whilst the World Health Organization recommends between 8 and 15 m^2 green areas per habitant. Local children may have not seen a dry forest, and statistically, most of their parents or ancestors migrated from the Andes; therefore, may not possess an ecological understanding and connection with this vulnerable and unique coastal ecosystem in which they now inhabit.

5.8 Project Partners

5.8.1 Conciencia

Peruvian Conciencia is an environmental education nonprofit established to develop, design, and conduct outdoor educational activities. The name means both "awareness" and "with science". While science is at the core of understanding nature, Conciencia works across disciplines to establish nature as an incomparable matrix on which a wide range of topics can be delivered. "A Forest for Ancon" provided one such matrix to a community of students, promoting the delivery of outdoor lessons, access to green space, and hands-on science-based learning.

Fig. 5.2 Ancon, Peru during the summer 2015. A hillside favela alongside the *Panamericana* highway in Peru, near the project location

5.8.2 Las Colonias School

Union of Social Assistance Projects (Union de Obras de Asistencia Social, in Spanish), is a NGO that houses and provides education for hundreds of children throughout Metropolitan Lima, in addition to managing assisted living facilities for the elderly. The facilities involved in this project are Las Colonias schools, which house and educate 400 children deemed vulnerable or at-risk. The property and management are divided into one boys' and one girls' boarding school, operated with autonomy by their corresponding managers and school principals. Each houses approximately 210 children. Most of the children come from single-parent families, often falling below the poverty line. Although the housing is operated by the non-profit, the schools are operated like any other public school in Peru, by the Ministry of Education.

Conciencia and Union of Social Assistance Works began their partnership in 2011, with Conciencia conducting environmental workshops and playful learning activities, including a public beach clean-up during a 2011 surf championship. At its inception the forest in Ancon was planted with 450 trees. The intention was to match the number of students (including an estimated 10% yearly tree mortality in the first 3 years).

5.8.3 Financial Partnership Through Corporate Social Responsibility

Conciencia partnered with a multinational company through their annual volunteer day. A full day activity was designed for their Lima-based workforce of just over 200 employees. The employees volunteered their time and helped dig holes and plant trees, supported by a coalition of forestry students and recent graduates from the Universidad Nacional Agraria La Molina (La Molina National Agrarian University).

5.8.4 Visiting Students

In the years following the establishment of the forest, Conciencia brought other students to the area. Conciencia prepared day-long educational activities for students participating in educational tours from overseas. Traveling students visited the area and conducted monitoring activities, guided by forestry engineers. Activities included measuring tree growth, assessing the presence of weeds, plant health and phenology, as well as observing the insects and birds, as shown in Table 5.1.

The non-profit also taught the students about the ecological history of the area and the broader region on site, allowing the students to see some of the endemic and endangered species.

Table 5.1 Students and volunteers track tree health and phenology indicators, using a guide to identify each young tree

Tree ID	Species	Height (m)	Phenelogy		Health		
			Fruit	Flower	Sick	Healthy	Dead
1	Algarrobo	0,8				x	
2	Tara	1,86				x	
3	Algarrobo	0,3			x		
4	Tara	2,5				x	
5	Algarrobo	0,23			x		
100	Molle	1,55		x		x	
101	Molle	1,52		x		x	
102	Huaranguay	1,38			x		
103	?						x
104	Molle	2,23		x		x	
105	Tara	1,83				x	

5.9 Designing an Agriculture Learning Project: Strategies and Consideration

As an organization, Conciencia is characterized by a design-thinking approach, guided by resources and training from organizations such as IDEO, a global design company (IDEO 2012) and the Stanford University school of design (https://dschool.stanford.edu/).

5.9.1 Human-Centered Design

Human-centered design is at the base of Conciencia's approach, due to the social nature of our work. Human-centered design approach considers projects should start with a high degree of immersive research to generate empathy for our users. Users are brought along on the co-design of programs, and designers should prototype quickly to gain further insight (Sorice 2015). While the primary users of this education project are the students, the secondary users are the teachers. This corresponds to two distinct outcomes: (1) the primary outcome – to build environmental literacy and identity, and (2) the secondary outcome – to incentivize teachers to use the green space to deliver lessons, through their own agency and initiative. The first and most crucial part of any project design is to understand the context.

In Ancon, the project designers found a gap in the development of environmental literacy and identity through place-based education. Interviews revealed that field trips were difficult to organize due to shortage of funding and infrastructure services, such as buses. This primed the partners to use the existing space to amplify the outdoor learning opportunities. One of the schools had a small farm on site and students were excited to interact with the animals.

Conciencia's mission is to discover the education potential in students' interests in nature. For this purpose, the preferred methods are to take students and educators into nature to host a lesson. Before, during and after the lesson, facilitators observe what captivates students' interests, what distracts them, and what offers risks. We used this information to tailor activities. This is a great way to develop multidisciplinary lessons that are fueled by the students' interests (Sobel 2008). We may use cues to incentivize students' explorations. David Sobel's book *Childhood and Nature: Design Principles for Educators* is a great resource for us. In his introduction, he arbitrarily separated recurrent play themes into seven principles he has observed in children in his long career as an environmental educator. These principles serve as guides to structure any type of place-based learning and they are effective in our experience.

On other occasions the topics naturally arise. Never doubt a child's cognitive skills to uncover profound nature processes insights. Some of our most brilliant lessons have been proposed by students: one may begin to charter down one line of investigation, as others chime in and suddenly a class-wide lesson is born. As Louise

Chawla, an influential childhood researcher, declared: "Perception is commonly shaped by processes of joint attention – when people attend to the same thing together" (Chawla 2006). In one of her prior studies, Chawla discovered that opportunities for action rather than passive learning prevails in school memories (1998).

5.9.2 Children and Trees: Fostering Environmental Literacy and Identity

The forestry project is a prime matrix for any type of learning. However, in order to achieve our main objective of enriching environmental literacy and identity, we want students to accomplish a connection with the individual plants. By patiently observing, children can easily learn to recognize trees as living, growing, changing beings. By recognizing that children and trees share these characteristics, they start to pave the way for empathy with a variety of living beings.

To create and foster an environmental literacy and identity between the children and the trees, we implemented the following steps: (1) accessibility, (2) identification, and (3) adopt a tree.

The first requirement to create an environmental identity is to offer accessibility to the environment. Therefore, we created a safe place where children can interact with nature.

Children need direct and meaningful experience with nature to promote affinity and subsequently identify with it (Davis et al. 2006; Kaplan and Kaplan 1989; Seel et al. 1993).

In our first activity, children were given clues regarding the tree's shape (flowers, leaves) and colors so they could identify and recognize the tree species by themselves. Elaine Brook, an outdoor-based educator, explains that people are unlikely to value what they cannot name: "One of my students told me that every time she learns the name of a plant, she feels as if she is meeting someone new. Giving a name to something is a way of knowing it" (Louv 2006). The children in Ancon learned that each tree has different characteristics which makes them unique from each other (e.g., shape, structure and disposition of the leaf; color of the flower). They also learned that these characteristics have a purpose, and their survival depends on them (e.g., deep root system to gather water in the desert). Through play and art-based learning activities children were encouraged to observe painstakingly for details and small differences. At the end of the activity, girls and boys look at the trees with a different view. They have an insight: each tree was unique.

The next activity was intended to help the children create a personal and emotional bond with the forest and through it create roots to their school. Children leave after 6th grade, given only primary school is offered. The intention is that this experience will be significant and transformative, creating long term impact for these children that may ripple down to the next generation. In *Wild Hope: The Transformative Power of Children Engaging with Nature,* Charles and Louv end their research review by concluding, "[w]hen children are transformed by the healing, generative, and abundant gifts of healthy natural environments, they are more likely to grow into adults who will create those healing and transformative opportunities for their own children and others" (Charles and Louv 2018).

Each child adopted one tree: choosing and naming it. Conciencia staff provided opportunities, tools, and supervision for the students to de-weed the tree, and trim them to enhance their development. Students monitored their tree's growth (Fig. 5.3), applying math skills. Children were enthusiastic and motivated with the connection they felt with the trees as fellow living, growing beings. Nothing describes more authentically this experience than the children remarks: "We will be the fairy godmothers of these trees! We can visit them when we are grandmothers with our families".

Fig. 5.3 The activity tracking the tree's and students' growth to foster empathy as part of the project A Forest for Ancon

5.9.3 Example Activity

One of the activities carried out with the students through the course of the semester is tracking the tree's growth alongside their own (Fig. 5.3). This helped children understand that humans and trees share qualities of growth, health, and nutrition. Through this simple monitoring exercise students could explore what factors contribute or hinder the growth rate and how to calculate averages.

5.9.4 Designing for Teachers

Nature is an excellent matrix for any study subject. This area provides teachers with a diversity of teaching opportunities. One could teach an art class based on the observation of nature, as well as one could teach a science class or math class, as pointed out in the previous section. However, there are many barriers that prevent schoolteachers from bringing children outdoors. One important factor is class numbers. A high student to teacher ratio prevents teachers from maintaining safe conditions when transporting classes outdoors. A learning curve is to be expected both in outdoor classroom management and lesson delivery through a new medium, as most teachers rely on educational approaches that they have experienced themselves (Ross and Bruce 2007).

A big challenge for *A Forest for Ancon* is to promote enthusiasm from the teachers to hold classes in these areas. The chosen strategy was to apply a human-centered design process, based on empathy: this means that the project designers were tasked with really understanding the teachers, so that the project could solve a problem for them. In this instance, the teachers must be centered as users and indicated by accepted design-thinking methodology (IDEO 2012).

Initial group workshops prior to the project design revealed important considerations. The number one reported concern or pain point for teachers was time scarcity. Teachers indicated interest in a learning project that met their existing curriculum objectives. They indicated resistance to projects that would detract from them meeting their curricular goals or would demand extra planning and coordinating time from them. Therefore, project designers focused on the specific objective of identifying trans-curriculum learning opportunities that could be facilitated in the forest, or at least referencing the space.

It was important and helpful for the designers to develop empathy for the teachers as users of the project. Shadowing teachers was the better way to gather insights than group workshops or meetings outside the teaching schedule. Shadowing, observing classes on a regular basis without intervening, allowed designers to understand classroom dynamics, curriculum delivery methods and student behavior patterns, while giving access to one-on-one time with teachers between classes and creating a stronger sense of familiarity. During these conversations, teachers expressed their thoughts, concerns and suggestions to a greater degree than during the workshops.

By shadowing and interviewing the teachers at Las Colonias school we learned that none of them had received environmental education training in the outdoors. Reported environmental education training included recycled handicrafts and use of traffic signals, but training did not venture into maximizing outdoor potential for interdisciplinary lessons. Studies indicate that many teachers feel ineffective teaching outside (Ferry 1995), and lack of training is a substantial barrier to outdoor nature-based teaching (Carrier 2009).

5.10 Prototyping: Trial and Error

Conciencia, as a small non-profit organization, is able to prototype fast and cheap. This is a commonly accepted strategy within design thinking methodology (IDEO 2012): Smaller organizations have the opportunity of innovating. In contrast, larger organizations, and the government in particular face many barriers to innovation but hold the capability to scale (Koh et al. 2012). Prototyping allows for the study of the possibilities, challenges, and flaws in the design. Design thinkers believe that the best research happens in practice by observing how users use or misuse a particular tool (IDEO 2012). Documented instances of urban agriculture education success can build the case for large scale adoption of urban agriculture education.

5.10.1 Teachers' Perspectives

Through surveys, and conversations, the following insights we documented:

- Many teachers were immigrants, and their memories were of faraway mountains in the Andes. Their identification with the desert coastal area was low.
- Over 75% of the teachers travelled an hour or more in public transport to attend their job. The long commute was a barrier to after school commitments.
- Weekly or bi-weekly meetings and training sessions with the Local Education Management Unit (UGEL for its acronym in Spanish, part of the Ministry of Education) are time consuming.
- Class sizes were above 24 students, with only one teacher.
- Teachers worry about managing their class outside and students getting hurt. They generally feel they would lose control and be ineffective when teaching in the outdoors.
- Teachers requested a defined, shaded seating area to gather and instruct the group.
- The teachers consulted had never attended outdoor education lessons in their youth and, although about half of them reported affinity with nature as children, they had never attended a school-hosted field trip as students.

5.11 Agriculture Education as Punishment and Therapy

A follow up study revealed that while teachers had not incorporated the area in their lesson delivery, the head of one of the schools, an avid gardener, nun Juana Lopez did use the area, taking students with behavioral issues to gardening the forest. As she explained: "It is a form of therapy. It calms them down". To avoid giving the impression to behaved students that misbehaved students were rewarded, Ms. Lopez framed the activity as a punishment. In fact, youth gardens can reduce stress, improve attitudes toward school, facilitate collaboration and teamwork and intercultural awareness, improve peer relations and prosocial behavior, and improve self-efficacy and self-esteem (Kim et al. 2014; Rogers 2018). Evaluations of garden programs indicate that involving youth in higher-order responsibilities such as garden planning and decision-making results in higher levels of participation and leadership development (Rogers 2018).

5.12 Conclusions and Looking Ahead

5.12.1 Our Recommendations for Future Designs in Urban Agriculture Education

5.12.2 Team Diversity

In *Last Child in the Woods*, Richard Louv interviewed several educators and concluded that nature is an infinite reservoir of information, where countless lessons can be discovered. The team that worked on this project at different stages was multidisciplinary in nature. This helped bring to the surface diverse learning possibilities. The team behind this project had representation from earth scientists, artists, educators, a producer, a geographer, a nun, a social worker, with each member offering a unique perspective and able to cover particular insights and strategies to overcome the challenges and to help solve issues. The diversity of the team was one of the richer aspects of the project.

5.12.2.1 Making the Project Financially Viable

Partnering up with a private sponsor was key to getting the project off the ground. Increasingly, corporations are collaborating with non-profit and public sectors by donating funds, goods, and services through their Social Responsibility projects, and volunteering their staff (as Citibank did in the Global Community Day).

Neutralizing their carbon footprint may also be used as an incentive to finance future urban agriculture projects.

Ongoing costs can be challenging, particularly in difficult terrain like this one, where maintenance is necessary. This project carved out financial resources for three years of personnel, equipment, and material maintenance. Tree adoption schemes are a proven method of financial resourcing in Peru. A good example is the project "Reforestamos por Naturaleza" of Peruvian non-profit "Conservamos por Naturaleza". Their mission is to support private and community-based conservation initiatives and to promote a sustainable lifestyle. Their project "Reforestamos por Naturaleza" raised 27,700 soles (roughly 9200 USD) as a result of 1117 adopted trees. Thousands of trees were planted through an initial investment which began in 2016, and the organization allows people to adopt trees through a donation of 50 soles. Half of these funds are used to support the campaign and monitor the trees, and the other half goes to the private conservation area where the tree is planted. Donors receive the GPS coordinates of their tree (Conservamos por Naturaleza 2020).

The greatest source of economic income came from student travel groups. The non-profit provided an experiential educational service, generating an income for the forest's maintenance and for the local community who provided hosting, catering, tours, and transport services.

5.12.2.2 Suggestions to Remove Barriers: Teachers' Perspective

Paul Roberts of the Children Learning with Nature Institute points out that while many training programs tell teachers about the educational potential of nature, very few actually take the teachers out to experience an outdoor lesson. To move forward it is essential that teachers are trained in all aspects of outdoor based education in order to develop the skills, knowledge, and attitudes required to feel engaged, comfortable, and capable of successfully implementing education in outdoor settings.

A significant barrier for outdoor education is safety. Educators in other countries, like New Zealand and Scotland, are deterred by the risks involved in outdoor education, which is related to the adult/pupil ratios (Nicol et al. 2007; Zink and Boyes 2006). Reducing the student to teacher ratio in outdoor settings also reduces the likelihood of risk, as the extra supervision can prevent children from wandering off or suffering injury. Understanding educators' beliefs, practices, and perceived barriers relating to outdoor education can inform efforts to better provide them with the tools and training necessary to succeed at using these natural areas in education (Ernst 2014).

5.13 Scaling Up Urban Agriculture Education: ANIA Case Study

An essential component for the development of children and young people is giving them a space of land where they can nurture life and biodiversity with affection, through which they naturally establish a relationship with Mother Earth. This generates well-being for themselves, other people and nature. The creation and sustainability of these areas, coined TiNi (for its acronym in Spanish, translated to Children's Lands), have been the core mission of the Association for Children and their Environment (ANIA for its acronym in Spanish). This relationship with nature, that is nurtured in TiNi gives children and young people the opportunity to exercise their ability to transform the world, be positive agents of change and be valued and recognized by society.

TiNis can be implemented in homes, schools, neighborhoods, and communities, in urban and rural areas and in various ecosystems. In schools, TiNi has the following objectives: (1) to incorporate Mother Earth as a teacher to help children develop empathy for life, (2) to institutionalize Mother Earth's classroom as part of the educational infrastructure, used by teachers as a pedagogical resource to implement learning sessions across curricular areas, and (3) to foster the development of an emotional bond with Mother Earth in students and other members of the educational community. The TiNi movement has been recognized by UNESCO as a good practice of education for sustainable development and has been adopted by the Ministry of Education of Peru and Ecuador to be implemented in public schools.

In order to advance the 17 Sustainable Development Goals of the United Nations (https://sdgs.un.org/goals) and ensure their sustainability, it is essential that the new generations develop an active empathy for life. This is defined as the ability to prioritize the common good through daily actions that generate well-being for oneself, other people and Mother Earth, our essential ally. Who better than her to teach us to value and care for her?

References

Ardoin, N. (2013). *Influencing conservation action: What research says about environmental literacy, behavior, and conservation results*. New York: National Audubon Society.

Beresford-Jones, D. G. (2011). *The lost woodlands of ancient Nasca. A case-study in ecological and cultural collapse*. Oxford: Oxford University Press.

Beresford-Jones, D. G., Arce, T. S., Whaley, O. Q., & Chepstow-Lusty, A. J. (2009a). The role of *Prosopis* in ecological and landscape change in the Samaca Basin, lower Ica Valley, south coast Peru from the early horizon to the late intermediate period. *Latin American Antiquity, 20*, 303–332.

Beresford-Jones, D., Lewis, H., & Boreham, S. (2009b). Linking cultural and environmental change in Peruvian prehistory: Geomorphological survey of the Samaca Basin, Lower Ica Valley, Peru. *Catena, 78*(3), 234–249.

Blair, D. (2009). The child in the garden: An evaluative review of the benefits of school gardening. *The Journal of Environmental Education, 40*, 15–38.
Blatt, E. N. (2013). Exploring environmental identity and behavioral change in an environmental science course. *Cultural Studies of Science Education, 8*(2), 467–488.
Calabrese-Barton, A., & Yang, K. (2000). The culture of power and science education: Learning from Miguel. *Journal of Research in Science Teaching, 37*(8), 871–889.
Carrier, S. J. (2009). The effects of outdoor science lessons with elementary school students on preservice teachers' self-efficacy. *Journal of Elementary Science Education, 21*(2), 35–48.
Carrier, S. J., Tugurian, L. P., & Thomson, M. M. (2013). Elementary science indoors and out: Teachers, time, and testing. *Research in Science Education, 43*(5), 2059–2083.
Charles, C., & Louv, R. (2018). Wild hope: The transformative power of children engaging with nature. In A. Cutter-Mackenzie, K. Malone, & E. Barratt Hacking (Eds.), *Research handbook on childhood nature* (pp. 1–21). Cham: Springer.
Charles, C., Keenleyside, K., Chapple, R., Kilburn, B., Salah van der Leest, P., Allen, D., Richardson, M., Giusti, M., Franklin, L., Harbrow, M., & Wilson, R. (2018). *Home to us all: How connecting with nature helps us care for ourselves and the Earth.* Children and Nature Network.
Chawla, L. (1998). Significant life experiences revisited. *The Journal of Environmental Education, 29*(3), 11–21.
Chawla, L. (2006). *Learning to love the natural world enough to protect it.* Trondheim: Norsk senter for barneforskning.
Chawla, L., & Cushing, D. F. (2007). Education for strategic environmental behavior. *Environmental Education Research, 13*(4), 437–452.
Chawla, L., & Derr, V. (2012). The development of conservation behaviors in childhood and youth. In S. Clayton (Ed.), *Oxford handbook of environmental and conservation psychology* (pp. 527–555). New York: Oxford University Press.
Clayton, S. D. (2013). *Identity and the natural environment: The psychological significance of nature*. Cambridge, Mass: MIT Press.
Cobern, W. W. (2000). *Everyday thoughts about nature: A worldview investigation of important concepts students use to make sense of nature with specific attention to science* (Vol. 9). Dordrecht: Kluwer Academic Publishers.
Conservamos por Naturaleza. (2020). *Reforestamos por Naturaleza recolectó más de 27 mil soles durante el 2019.* https://www.conservamospornaturaleza.org/noticia/reforestamos-2019/
Davis, B., Rea, T., & Waite, S. (2006). The special nature of the outdoors: Its contribution to the education of children aged 3–11. *Journal of Outdoor and Environmental Education, 10*(2), 3–12.
Díaz, C. A. (1995). *Los Algarrobos*. Lima: CONCYTEC.
Ernst, J. (2014). Early childhood educators' use of natural outdoor settings as learning environments: An exploratory study of beliefs, practices, and barriers. *Environmental Education Research, 20*(6), 735–752.
Ferry, B. (1995). Enhancing environmental experiences through effective partnerships among teacher educators, field study centers, and schools. *The Journal of Experimental Education, 18*, 133–137.
Guiney, M. S., & Oberhauser, K. S. (2009). Conservation volunteers' connection to nature. *Ecopsychology, 1*(4), 187–197.
Hirschi, J. S. (2015). *Ripe for change: Garden-based learning in schools.* Cambridge, MA: Harvard Education Press.
IDEO. (2012). *Design thinking for educators.* http://designthinkingforeducators.com/
INEI – National Institute of Statistics and Information (INEI, for its initials in Spanish). (2015). http://www.inei.gob.pe/
International Organization for Migration. (2015). *Migraciones internas en el Perú.* Peru: Lima.
Kaplan, R., & Kaplan, S. (1989). *The experience of nature.* New York: Cambridge University Press.

Kim, S. S., Park, S. A., & Son, K. C. (2014). Improving peer relations of elementary school students through a school gardening program. *HortTechnology, 24*, 181–187.
Koh, H., Karamchandani, A., & Katz, R. (2012). *From blueprint to scale. The case for philanthropy in impact investing*. Acumen: Monitor Group.
Kuo, F. E. (2001). Coping with poverty: Impacts of environment and attention in the inner city. *Environment and Behavior, 33*(1), 5–34.
Louv, R. (2006). *Last child in the woods: Saving our children from nature-deficit disorder*. Chapel Hill: Algonquin Books of Chapel Hill.
Nicol, R., Higgins, P., Ross, H., & Mannion, G. (2007). *Outdoor education in Scotland: A summary of recent research*. Perth & Glasgow: Scottish Natural Heritage & Learning and Teaching Scotland.
Olivos, P., & Clayton, S. (2017). Self, nature and Well-being: Sense of connectedness and environmental identity for quality of life. In G. Fleury-Bahi, E. Pol, & O. Navarro (Eds.), *Handbook of environmental psychology and quality of life research* (pp. 107–126). Cham: Springer.
Osborne, J., Simon, S., & Collins, S. (2003). Attitudes towards science: A review of the literature and its implications. *International Journal of Science Education, 25*(9), 1049–1079.
Payne, P. (2001). Identity and environmental education. *Environmental Education Research, 7*(1), 67–88.
Project Dirt. (2018). *The impact of outdoor learning and playtime at school – and beyond: A summary of the survey findings conducted for Outdoor Classroom Day 2018*. https://outdoorclassroomday.com/wpcontent/uploads/2018/05/FINAL-Project-Dirt-Survey-Outdoor-Play-and-Learning-at-School-2018-15.05.18.pdf
Robinson, C. W. (2016). Children and nature. In T. M. Waliczek & J. M. Zajicek (Eds.), *Urban horticulture* (pp. 19–59). Boca Raton: CRC Press.
Rogers, M. A. (2018). Urban agriculture as a tool for horticultural education and youth development. In D. Nandwani (Ed.), *Urban horticulture* (pp. 211–232). Cham: Springer.
Ross, J., & Bruce, C. (2007). Professional development effects on teacher efficacy: Results of a randomized field trial. *The Journal of Educational Research, 101*(1), 50–60.
Seel, H. J., Sichler, R., & Fischerlehner, B. (Eds.). (1993). *Mensch — Natur*. Opladen: Westdeutscher Verlag.
Sobel, D. (2008). *Childhood and nature: Design principles for educators*. Portland: Stenhouse Publishers.
Stone, E. L., & Kalisz, P. J. (1991). On the maximum extent of tree roots. *Forest Ecology and Management, 46*(1–2), 59–102.
Tam, K. P. (2013). Concepts and measures related to connection to nature: Similarities and differences. *Journal of Environmental Psychology, 34*, 64–78.
Thomson, S., & Fleming, N. (2004). *Examining the evidence: Science achievement in Australian schools in TIMSS 2002* (TIMSS Australian monograph no. 6). Melbourne: Australian Council for Educational Research.
Tugurian, L. P., & Carrier, S. J. (2017). Children's environmental identity and the elementary science classroom. *The Journal of Environmental Education, 48*(3), 143–153.
United Nations. (2017). *World urbanization prospects: The 2018 revision*. New York: United Nations, Department of Economic and Social Affairs, Population Division. https://population.un.org/wup/.
Volk, T. L., & Cheak, M. J. (2003). The effects of an environmental education program on students, parents, and community. *The Journal of Environmental Education, 34*(4), 12–25.
Williams, D. R., & Dixon, P. S. (2013). Impact of garden-based learning on academic outcomes in schools: Synthesis of research between 1990 and 2010. *Review of Educational Research, 83*, 211–235.
Wolsko, C., & Lindberg, K. (2013). Experiencing connection with nature: The matrix of psychological wellbeing, mindfulness, and outdoor recreation. *Ecopsychology, 5*(2), 80–91.
World Wildlife Fund. (2014). *Sechura Desert*. http://www.eoearth.org/view/article/155957

Wortham, S. (2006). *Learning identity: The joint emergence of social identification and academic learning*. New York: Cambridge University Press.

Zink, R., & Boyes, M. (2006). The nature and scope of outdoor education in New Zealand schools. *Journal of Outdoor and Environmental Education, 10*(1), 11–21.

Chapter 6
Permaculture in Action: Urban Farming as Continual Science Learning

Zev H. S. Friedman and Phyllis Katz

Abstract This chapter tells a story of the Continual Science Learning (CSL, also known as informal science education, ISE) that both participants and educators experienced in two consecutive years of an intensive hands-on urban *Permaculture In Action* (PIA) program. The program took place in a mid-size southern Appalachian city in the U.S. We make a distinction between urban agriculture and Permaculture. The latter holistically combines food production, energy systems, the built environment, and social/economic considerations. We show how this program was structured using Permaculture principles to facilitate Transformative Permaculture Learning (TSL). We present evidence as to how we followed those principles to meet most of those goals. We describe how the curriculum cycle enacted the six strands described in Learning in Informal Environments (Bell P, Lewenstein B, Shouse AW, Feder MA, Learning science in informal environments, people, places, and pursuits. National Academies Press, Washington, DC, 2009), and how the Permaculture curriculum categories met major science, technology, engineering and mathematics (STEM) learning objectives. We describe how the evidence altered our teaching. We finish by discussing the broader land access, economic and societal dynamics that affect Permaculture educators' ability to create the kind of change needed to continue healthy human ecosystems on our planet. We also identify specific challenges and continuing research questions to explore.

Keywords Transformative Permaculture learning (TSL) · Permaculture · Permaculture education · Urban permaculture teaching · Urban agriculture · Informal science and agriculture · Continual science learning and agriculture · Continual science learning (CSL) · Science teaching in urban agriculture

"The environment is everything that isn't me."—Albert Einstein

---

Z. H. S. Friedman · P. Katz (✉)
Living Systems Design, Bainbridge Island, WA, USA
e-mail: zev@livingsystemsdesign.net

I. DeCoito et al. (eds.), *Teaching and Learning in Urban Agricultural Community Contexts*, Urban Agriculture,
https://doi.org/10.1007/978-3-030-72888-5_6

"To recognize our actions and ourselves as a part of nature is a cultural transformation, begun but not completed.... We must rebuild and surpass the long-term thinking that was natural to indigenous cultures of place. Learning how to see and feel the multi-dimensional, patterned nature of the earth's living history is central to the ability to use and creatively respond to the change we face" (David Holmgren 2002, p. 265)

6.1 Introduction: Permaculture Teaching as Human Adaptation

This chapter discusses one example of an effort to be rebuilders and long-term thinkers through Permaculture teaching and land transformation. Permaculture is a leading example of applied systems thinking connecting the dots between polluting "waste" products and ways to use those wastes as food to feed other processes, reconnecting fragmented communities for mutual benefit, and reframing the human role in the planetary biome as that of the wise healer instead of the short-sighted extractor (Mollison and Holmgren 1978). We have used the lens of what has been called informal science education (ISE), but which we refer to as continual science learning (CSL) – what happens when science education is needed after compulsory schooling has been completed (Katz 2017). While Einstein, in the above quote, seems to have meant a broad sweep of inclusion, by setting himself aside, he reinforced the notion that people are outside the system. As the human footprint increases, we must realize how our numbers, technology, and creations are changing the world on which we depend. There is diverse evidence that those changes are damaging the ecological fabric of the world on which we depend (e.g., Darimont et al. 2015; Palumbi 2001; Vitousek et al. 1997). Permaculture is one way in which we can use our human capacity to engage in CSL to alter our behaviors – to adapt – to sustain the earth.

CSL practices and research enable us to continually adapt. Human adaptation occurs within communities through social policies and practices. Every decision impacts a future set of choices. We largely enculturate our policies and practices in our school science curricula where we select and teach what we want our students to learn about the world as we understand it and as we would like it to be (Next Generation Science Standards [NGSS] 2013). When we are not in school we learn science from our personal experiences and from media. Science education beyond schooling is vital as science research continues to produce relevant information throughout our lives (e.g., Bell et al. 2009; The Royal Society 1985).

We were interested in studying the CSL teaching and learning that occurred in this particular setting of a voluntary adult Permaculture program. We were interested in whether and how our teaching design functioned *in situ*. We have found a number of articles on science education through Permaculture (Anderson et al. 2019; Arko-Achemfuor 2014; Hockin-Grant and Yasue 2017; Lebo III 2011; Mobus 2018; Praetorius 2006; Rios 2010). It has been difficult to assess how much and what kind of hands-on, in-depth Permaculture teaching is occurring around the world. Often, those who practice Permaculture are not academics seeking to publish. The most centralized source for Permaculture-related research is the Permaculture Research Institute in Australia (http://Permaculturenews.org/). As of October 2020, the website lists 2663 Permaculture projects worldwide. The list and map can be filtered by leader, climate, and country. The website invites self-report and does not have a specific search category for educational research. It is our overall aim to contribute to Permaculture education by describing this case study of Permaculture as continued adult science teaching and learning.

We approach our work using both Permaculture theory and continual science learning (CSL) theory. For the reasons stated above, we believe it is critical to provide science education opportunities in land use that acknowledge the earth as a recycling system in which we (humans) must consider and act upon our impact within that system. We have prepared this chapter to give the reader the theoretical background of Permaculture, the CSL theory that is intertwined with the educational design of this particular effort, and the challenges of teaching this critical earth stewardship to interested adults. We offer not only our specific outcomes, but suggestions for science educators in and out of schools who engage in land use projects (and Permaculture especially) in their teaching.

6.2 Permaculture Teaching as Continual Science Learning

"Permaculture" is a term derived from the words "permanent" and "agriculture/culture", introduced by Australian researchers Bill Mollison and David Holmgren (1978). Mollison and Holmgren studied existing and historical cultures around the world to find commonalities in those cultures where people lived in one place for many centuries without undermining the ecological integrity of their landscape. They found remarkable commonalities, combined them with modern concepts and information from ecology, agriculture, horticulture, systems thinking, geology, soil science, hydrology and other fields, and created the concept of Permaculture. Permaculture expresses a vision of a type of agriculture, economy and society that would be more resilient than the "bubble" model of growth apparent in industrial farming and economies. In the bubble model, rapid unsupportable growth occurs via non-renewable practices, with a predictable collapse as the endpoint. In practice, Permaculture has developed as a pragmatic design system for creating regenerative human habitats and cultural lifeways at all scales. Regenerative projects create systems now that actually enrich the basis for human welfare and ecological vitality for future generations. This standard differs from "sustainability" which seeks to mitigate environmental problems and is often defined as "meeting the needs of the present without compromising the ability of future generations to meet their own needs" (United Nations Department of Economic and Social Affairs 1987).

Using insights from natural ecosystem dynamics via a set of design principles and mentor-based training, Permaculture practitioners apply an interdisciplinary mix of knowledge and techniques to create a phased plan for a given land project, business, community or other endeavor that will meet human needs while healing damaged ecosystems. There are different versions of the Permaculture design framework, all of which are derived from natural ecosystem patterns and applied to land use, organization, and other human endeavors. For example, just as a nut tree in a forest ecosystem performs many different functions via complex food web interactions (habitat for squirrels, food for many animals, soil creation and erosion control, feeding fungi, sequestering carbon, water retention and nutrient cycling, providing lumber and so on), we design elements in Permaculture systems with the key principle that *each element performs multiple functions*. The Permaculture in Action (PIA) program that this chapter describes follows this principle and was designed to perform at least 13 functions as described in the goals section below (Marsh, Allison and Friedman 2014). Another Permaculture design principle, *consider succession,* reminds designers to learn from the process of ecological succession; we can thereby design land use systems and organizations to mimic the phased growth and transformation of strategies through time that allows ecosystems to resiliently respond to change, maintain species diversity, and retain nutrients and topsoil.

6.3 Urban Permaculture Teaching

In the past 10 years, urban Permaculture has become much more common, and in many ways, a primary focus of the global Permaculture effort. Globally, Permaculture has significantly impacted urban areas, where any attempt to address the related challenges of poverty, food, institutionalized racism and pervasive environmental toxins requires the kind of interdisciplinary design approach offered by Permaculture. It makes sense – cities are where most of the people are – 55% worldwide (Ritchie 2018). There are other considerations, for example, many marginalized and low income people are concentrated in cities and can benefit the most from growing their own food in the city. Ideas can spread quickly in cities, so they are high leverage places to act to affect climate change and other issues. Space is already tightly managed in cities (lawns, landscapes, buildings) facilitating reconsideration of land management, with policy changes toward utilizing that space for beneficial yields while cutting fossil fuel energy use.

There is a distinction to be made between urban agriculture and urban Permaculture. Although growing annual vegetables in cities is usually an improvement over typical twentieth century urban land use, a core insight of Permaculture is that integrated agroecology farming systems that imitate natural ecosystems by including animals, fungi, annual plants, perennial plants and woody plants are much more stable and truly regenerative than annual farming. These Permaculture systems operate with fewer inputs and offer higher yields per land unit, and they offer novel solutions to urban challenges that cannot be achieved by organic row cropping techniques.

In one project in Butula, Kenya, they were especially effective in a high poverty village (Hockin-Grant and Yasue 2017). In our setting, in one project we completed, laying chickens were successfully used in a concentrated corridor (a "chicken moat") along our client's property line to prevent the encroachment of invasive plant species (Japanese Knotweed and Japanese Honeysuckle) from a nearby public property. The chickens' natural behavior is to eat fresh growing plant tips and scratch at root systems to forage insects and latent seeds. Locating them where invasive plants are an issue allows them to "patrol" that area, eating weeds and reducing the human labor needed for maintenance. Inside the eight foot wide chicken moat the ground consists of hardwood woodchips inoculated with Winecap Stropharia mushrooms (Stropharia rugosoannulata), an edible mushroom species whose fine hyphal threads create a spongy "myco-filter" in the woodchips. This arrangement catches and destroys fecal coliform and E. coli bacterial species that might be present in the chicken manure, thereby releasing cleaner stormwater into the landscape and ultimately into the urban stormwater system (other mushroom species can be used to create myco-filters that catch and metabolize hydrocarbon pollutants such as motor oil and herbicides (Stamets 2005)). The stropharia mycelia also create food and habitat for additional insects, who in turn feed the chickens, thus reducing purchased chicken feed in the warm season. Also, in the chicken run are mulberries and elderberries which are watered by rainwater from the roof of a nearby building via "pit-swales", shallow infiltration pits filled with woodchips and located uphill of each shrub. These large shrubs are fertilized directly by the chickens and provide food and anti-viral medicine for the landowners as well as high quality chicken forage and summer shade for the birds. Concentrated manure from the chickens' night-time housing is harvested and composted as a soil amendment and fertilizer for vegetable production.

The aforementioned system integrates animals, trees and shrubs, fungi, and annual vegetables in a designed food web that yields many valuable foods and medicines, limits inputs, cleans stormwater, and balances invasive plant dynamics. It requires more knowledge to establish than row cropping of annuals. Management of this system is more complex than the latter, but it accomplishes many vital functions which we believe are essential in the evolution of the healthy cities that we need. It is teaching for such changes, or transformations, that requires new vision and reorganization of roles in education (Anderson et al. 2019).

Since we have described Permaculture as a pragmatic design system for creating regenerative human habitats, it directly connects the lives of students to applied science, technology, engineering and mathematics (STEM) education. Lebo III (2011) makes the case for identifying those who practice Permaculture as "citizen scientists," noting that such people use both scientific content and process for "science in action". Learning how to manage land regeneratively requires the continuing science learning that happens after formal schooling has been completed and as needs arise. The principles of science learning presented by those who work in out-of-school learning and published by the U.S. National Research Council, are listed as:

- Knowledge, practice and science learning commence early in life, continue throughout the life span, and are inherently cultural.
- Science is a system of acquiring knowledge through systematic observation and experimentation.
- The body of scientific knowledge that has been established is continually being extended, refined and revised by the community of scientists.
- Science and scientific practices weave together content and process features.
- Effective science education reflects the ways in which scientists actually work (Bell et al. 2009, p. 42).

There are six strands of how learners engage that include goals and practices (Table 6.1).

We have seen how this PIA Permaculture teaching program and its philosophy, theory, and goals are a CSL exemplar of need-motivated learning beyond school.

6.4 Context of this Setting

This chapter uses data from Permaculture teaching programs that took place in the mid Appalachian region of the southeastern United States in 2012 and 2013. The participants attended five weekend sets of classes over two and a half months in 2012 and seven weekends over five months in 2013. The program was designed to introduce and support Permaculture principles and their practice. About two-thirds of the program involved hands-on practice and about a third was devoted to readings and discussion.

Table 6.1 Six Strands of ISE Learner Engagement (Bell et al. 2009, p. 43)

Informal Science Education Strands					
1: Experience excitement, interest and motivation to learn about phenomena in the natural and physical world.	2: Come to generate, understand, remember and use concepts, explanations, arguments, models and facts related to science.	3: Manipulate, test, explore, predict, question, observe and make sense of the natural and physical world.	4: Reflect on science as a way of knowing; on processes, concepts, and institutions of science; and on their own process of learning about phenomena.	5: Participate in scientific activities and learning practices with others, using scientific language and tools.	6: Think about themselves as science learners and develop an identity as someone who knows about, uses and sometimes contributes to science.

6.5 Participants

Students were recruited via PIA website, regional networking, informal contact, fliers, conferences, educational events, and social media. In 2012, most of the students were Caucasian, but some self-identified as Cherokee, Asian or African American. There were seven men and nine women in this group. Two had Master's degrees, 10 had Bachelor of Arts degrees, one was in nursing school, one had "some college", and two did not provide this information. In 2013, there were 12 women and seven men. Fourteen self-identified as Caucasian, and the others as Native American, Hispanic, and East Indian. Eleven had Bachelor of Arts degrees. One had a law degree, while others had some college or specialized education.

The program was fee supported as are many CSL programs. The cost was an impediment for some students and the designers/teachers worked out a method to address this obstacle. Since ongoing Permaculture properties were the sites for the practice, the fee was reduced by an equation that took into account the labor being provided for the landowner. The students were also interested in a higher proportion of hands-on to classroom style learning. Through learning while doing work that needed to be done, the Permaculture teaching was implemented in the whole systems design context. That is, this arrangement was consistent with the philosophy of integrating needs within a project. The teachers and students needed sites for experience and the landowners wanted work done which would transform their properties into self-perpetuating places. This mutual need has been reported in other projects (Arko-Achemfuor 2014; Hockin-Grant and Yasue 2017). We believe that reducing program fees in exchange for labor makes the approach more accessible and sustainable. Students were divided into teams so that the work could simultaneously fulfill the landowner requests. There were nine work sites each year. Due to space limitations, we include below a sample of the landowners' data for the reader to view site variation and scope of projects (Table 6.2).

Zev, the lead teacher and co-author of this chapter is a Permaculture practitioner, designer, researcher and instructor in the southern Appalachian region. He received his Bachelor of Science degree in Human Ecology from UNC-Asheville in 2004 and his Permaculture Design Certificate in 2007. He then completed a 2-year Permaculture teaching apprenticeship. Zev has 12 years of experience in urban Permaculture and 8 years in professional design and installation. His concern for quality Permaculture teaching is a component of our work and this chapter. Dylan Ryals-Hamilton of Transition Asheville (helping to plan for western North Carolina's future as an abundant, self-reliant network of towns, cities and rural areas in this age of petroleum decline, climate change, and global economic instability) co-organized and co-taught for the program.

Table 6.2 Sample of Landowners' and Permaculture Teaching Activities

Property/Owner	Permaculture Teaching Activities
Private property 6, 1/2 day This woman and her partner live in a small house on 1/3 acre in the city's downtown area. She has a strong commitment to community scale Permaculture and neighborhood collaboration, and her property is on a prominent street corner which has attracted lots of attention ever since her Permaculture transformation began in 2009. She was part of the organizational and promotional team for PIA and we traded her 1/2 day of work at her property as part of her compensation package.	1) Tour and overview of Permaculture design for property. 2) Planting of Rosa Rugosa/ comfrey "fedge" (food hedge) along property line with sidewalk. 3) Digging and creation of 25′ × 6′ swale pond, approximately 1500 gallons, at high point in property's topography for rainfed crop irrigation. 4) Digging and shaping of contoured "milpa" cropping beds designed for flood irrigation. 5) Sheet mulching and cover-cropping to reclaim bermuda grass inhabited areas for planting. 6) Planting blueberry/goumi guild in swaled south-facing roadside area.
Church property, 2 full days This church is a large structure with a very active program serving the homeless population of the city. In addition to multiple programs helping homeless people to gain power in their own lives, they regularly serve dinner to 500 people each week on Wednesday evenings, and anyone in the community is invited to attend. This is a place where people who don't often cross paths can meet and get to know each other. We chose this location as our community-scale service project for the class because some previous students had developed a relationship with the leaders of the congregation and completed a Permaculture design for the 3-acre property as part of our Permaculture Design Certification class the previous year.	1) Permaculture principle review. 2) Two hour forestry presentation, open to public as well as PIA students. 3) Students present Permaculture master plan for property. 4) Work project installing contoured vegetable production beds and weed control measures on slope adjacent to existing garden. 5) Sheet mulching large perimeter area of property. 6) Community-scale Permaculture panel and discussion open to public (150 class participants/ community members in attendance).

(continued)

Table 6.2 (continued)

Property/Owner	Permaculture Teaching Activities
Private property 4 (Repeat landowner from 2012) 2 days Both in early 60s, 2/3 acre property in East Asheville, heavily involved in Transition Asheville initiative, wanted to create a demonstration of highly productive urban Permaculture for Transition community to participate in and learn from.	1) Introduction to Permaculture design for property. 2) Construction of second, 500-gallon pond that overflows from first rain fed pond and serves as swimming and drinking water for ducks and creating fertigation for forest garden in near vicinity and downhill. 3) Planting of pawpaw guild. 4) Planting elderberries and Pakistani mulberries. 5) Planting living black locust/willow arbor for hardy kiwis, native wine grapes. 6) Inoculation of stropharia woodchip bed under Nikita's Gift persimmon within primary duck paddock. 7) Pollarding of existing 8-year old black locust trees as nurse trees for mulberry, butternut, persimmon, pawpaw. 8) Installation of additional duck paddock fencing corridor.

6.6 Integrated Principles and Program

6.6.1 Goals and Methods for Teaching

This section first describes the Permaculture educational goals and then those that are more operational, related to teaching within the context. We reference the CSL learning strands. Permaculture allows practitioners in any region to plan integrated land use systems which meet a community's needs. Strand 1: Motivation, includes income, food production, water systems, carbon-neutral electricity, energy efficient building materials and the recycling of waste, as well as social needs such as youth education and elder care. This training requires hands-on learning before participants can successfully implement projects, Strand 3: Manipulate. Successful learning during the implementation of Permaculture projects relies on experiential teaching by the leaders. For these reasons, we created a unique format for in-depth, hands-on training for adults who want to design and implement urban Permaculture systems, Strand 2: generate and use concepts; and Strand 5: Participate in scientific practices.

We planned a program with clear goals from which we could learn and that could accomplish multiple simultaneous functions consistent with Permaculture itself. Permaculture is built upon an ethically based design system. The ethics are commonly articulated as *care for earth, care for people, and fair share of resources.* Our prime concern was to offer a learning experience that instilled the ethics of Permaculture and the core value of providing regeneratively for future generations. We frequently referenced Permaculture ethics in the introductory section of class each morning and as a basis for decision making in presentations of Permaculture master plans for the landowners with whom we worked. We sought to model the ethics by demonstrating how each Permaculture plan increased ecological diversity and health, decreased the landowner's resource use and negative impacts, and provided long-term ecosystem benefits including carbon sequestration. People care was built in through simple practices such as periodic stretching, water breaks, finding ways for people to participate who had different body abilities, teaching ergonomic work processes and safe tool use, and demonstrating a culture of respecting and taking care of each other when injured, sick or going through difficult life processes. Fair sharing of resources was modeled through making the class tuition affordable for students and work contracts affordable for landowners, offering partial work-trade positions to attend the class, giving away plant propagates and seeds to students and offering a free Community Scale Permaculture Panel attended by 150 people.

The primary organizing principle of the content was to teach actionable, hands-on skills to participants in an explicit context of whole systems thinking and design (considering the relationships and mutual impacts within a system). Each morning of class began with a review of Permaculture design principles and different exercises asking students to apply these whole systems principles to their personal lives, finances, communities, landscapes, relationships or businesses. We then reviewed the Permaculture master plan drawing of the property we were working at that day so participants could understand the client's goals and the interconnected system we designed to meet those goals. Finally, we described the installation projects we would be working on that day, why we were doing these particular projects in the timeline and context of the master plan, and discuss nuances, physical details, tools and materials involved in those projects, before grouping into squads of four to six students each led by an apprentice crew leader to work on the projects (Strand 5: Participate).

We also wanted to educate participants and the community at large about the broader application of Permaculture design beyond its most publicized use as a gardening approach, emphasizing the application of ecological design principles to social and cultural systems. We accomplished this goal mainly through the morning Permaculture principle activities. Each morning we asked students to apply these ecologically based insights to different realms of human endeavor and needs, thereby highlighting the analogous dynamics between intrapersonal, interpersonal, community scale, economic and ecological systems (Strand 6: Identity as science learner and user). For example, the Permaculture principle "Produce No Waste: Waste is A Verb, Not A Noun" is demonstrated in biological systems including at the interface between our own bodies and the ecosystems we live in, where "waste products" that our bodies produce (urine and feces) easily compost down into high quality nutrient sources for plants. But if we treat human manure as a waste product to be disposed of instead of safely composted for use, it becomes a problem that requires vast amounts of water, money, resources and human energy to "treat" via urban municipal waste treatment systems which have major costs and flaws. This same dynamic occurs in society where groups of (usually very low income) people are marginalized, isolated and treated as "waste" products by the industrial economy and institutionalized prejudice. Because our society does not recognize inherent value in all groups, it does not work to create ways for them to meaningfully contribute. This dynamic breeds a downward spiral that destroys trust and in which marginalized groups become decreasingly powerful, they have less access to resources and information, and desperate actions such as violent crime are perceived as the only option to meet basic needs.

Permaculture thinking will seek to find gainful ways to include, learn from and collaborate with all interested individual and groups so harmful dynamics are transformed into healing and empowering processes. For example, a local organization works with ex-inmates to employ them with living wage jobs to run community gardens in the neighborhoods in which they committed their crimes. This strategy gives them a chance to make income, break the prison cycle, mend relationships in their communities, gain access to good food for themselves and their communities, and learn science and marketable skills. Another strategy for linking the application of Permaculture principles to a larger audience was the Community Scale Permaculture Panel. In 2013 this was a half-day event where an elder Permaculture instructor, an expert in grassroots organizing and decision making, a City Council member, a young African American mother and an Environmental Policy professor from a nearby university served on a panel to discuss practical steps for transforming a diverse city using Permaculture thinking.

Learning from the British based Transition Town initiative (Wells 2011), we also designed the class to engage everybody who was involved, including students, landowners, teachers, panelists, apprentices and the public, through the integration of "head, heart, and hands", i.e., intellectual engagement, emotional involvement and physical action. Recent brain research has shown that emotions are well integrated (and not separate from) the learning process (Immordino-Yang and Damasio 2007). By its nature, CSL (non-compulsory) learning can only happen if participants gain emotional as well as intellectual satisfaction from their experience. Emotional connection is, then, a necessary component in CSL in terms of attraction and retention, and aligned with Strand 1: Motivation, excitement. Fischer and his colleagues have brought "head, heart, and hands" together in dynamic skill theory (Fischer and Rose 1998; Mascolo and Fischer 2010; Yan and Fischer 2002). This approach begins with who we are as humans, through the insights of evolutionary biology and builds upon the work of Piaget and Vygotsky and their contributions to our understanding of development and the social interactions necessary to human learning. Most importantly, as an integrated systems approach for individuals, it is a good match to the systems thinking approach that is an underlying philosophy and practice of this adult Permaculture program. Insights into human learning as a dynamic system open the way to better understanding of the dynamic systems of the earth itself and our place in it. We would like to explore this connection further.

Looking toward the long-term goal of how the program impacts participants, we wanted to empower everyone involved to follow through with fundamental lifestyle and behavioral changes supported by a community network among students, landowners, apprentices, and teachers. The last day of the class was devoted to a collection of activities called "Where do we go from here?" This section was designed to bring resolution to the learning experience, encourage introspection on what was learned and the value of the whole experience and provide opportunities for participants to express hopes, fears and needs they had around implementing this learning in their lives, aligned with Strand 4: Reflect on science and their own learning. We shared book, website, video and other references, as well as contact information. We gathered feedback on how to make the program better in the future. We also then requested permission to add students to our mailing list so they would be in the loop for future events and learning opportunities – a modern CSL marketing tool.

In addition to the educational content goals described above, we developed testable operational goals which helped us evaluate our experience as Permaculture teachers, as listed below.

1. To make the class affordable for participants and the installation work affordable for landowners.
2. To utilize human energy to establish living examples of high-quality urban Permaculture demonstration sites for testing techniques, giving tours, teaching other classes in the future. Here we contrast "human energy" to "fossil fuel energy".
3. To document the class extensively via video and photo footage to be used for instructional materials and for future promotional use.
4. To gather feedback and learning data from students to determine efficacy of teaching methods and to improve/promote future classes.
5. To design this class so it is a prerequisite for later, more advanced opportunities with us and other instructors.
6. To develop a unique set of teaching skills and a niche in the Permaculture and agro-ecology learning arena.
7. To identify and train leaders who can start projects and spread the knowledge after the class is over; in particular, to identify potential apprentices.

6.7 The Program Teaching Plan—Community Science

The course had ten major components within the theme of ecological systems design:

1. Introduction to Permaculture;

2. Ecological design process and methods;
3. Climate, water and biogeography;
4. Soil health, carbon sequestration and solid waste remediation;
5. Regional ecosystem;
6. Trees, forests and forest agriculture;
7. Plant use strategies;
8. Animals in our systems;
9. The built environment; and
10. Invisible infrastructure, technology and economics.

Students participated in the entire process from theory, to design, to implementation, to evaluation. They needed to learn STEM concepts for their projects. Table 6.3 provides a sample of the course sections, goals, activities, and STEM learning for four of the 10 components.

6.8 Data Collection and Analysis

For the operational goals, we documented our fee arrangements, monitored the labor hours, and captured visuals (photographs and videos). We also monitored the level of the class design and the apprentice recruitment yield. In 2012 and 2013 we conducted student interviews for feedback on their learning and our teaching. Sample questions included:

Why did you sign up for this program?
What science or other knowledge do you think you will acquire in this program that will allow you to practice Permaculture in your life?
Please tell us your current best definition of Permaculture?
What are your strongest values in life and how are they connected to your everyday lifestyle?

We reviewed these for common themes at the preliminary stage.

6.9 Findings

We are able to report that the operational goals were met. The fee arrangements discussed earlier in the chapter enabled those who wanted to be in the program to participate. The human labor in our program accomplished the Permaculture installations with fewer fossil fuels. Extensive photo and video documentation occurred. We successfully trained five apprentice crew leaders in 2012/2013, and each of these apprentices has now taken this learning into creative income endeavors: two are now in childhood ecological education, one is organizing Permaculture oriented music festivals, and two are professional Permaculture designers and installers.

Table 6.3 Permaculture in Action: Program (Sample) Summary

Course Components/ Design Principles and Goals	Content	Activities	STEM (NGSS Dimensions)
Introduction to Permaculture 1) Instill ethics of Permaculture 2) Teach skills in context of whole systems thinking and design 3) Application of ecological design principles to social and cultural systems 4) Inspire through head/ heart/hand integration	Ecological snapshot, history and definition of Permaculture, ethics and design principles. History of ecological movements, agricultural failures leading to civilization collapse; creation of Permaculture design principles; observation and information gathering techniques	Opening fire, group discussion of ecological challenges; lecture on Permaculture origination topics; 4-hour detailed introduction to Permaculture principles. Daily morning exercise asking students to apply Permaculture principle insights to diverse life arenas.	***Scientific practices***: Observation; Data collection; Discussion ***Content:*** Science in society; Technology development ***Crosscutting:*** Cause and effect; Systems; Stability and change
Climate, water and biogeography 1) Ethics 2) Instill ethics of Permaculture 3) Teach skills in context of whole systems thinking and design 4) Inspire through head/ heart/hand integration	Introduction to climate science, global carbon and nutrient cycles, hydrologic cycle. Micro-climates, water, wastewater and aquaculture, earthworks, befriending water in the urban Permaculture system. Cisterns, tanks, greywater systems, composting toilets. Building ponds, roads, swales, terraces and other earthworks…	Fireside lecture on carbon cycle; content delivered in morning design presentation, during installation of this infrastructure and during end-of-day walkabouts.	***Scientific practices***: Collaboration; Discussion; Tool use ***Content:*** Earth science: Hydrology; Technology; Engineering ***Crosscutting:*** Systems, structure and function; scale, proportion and quantity

(continued)

Table 6.3 (continued)

Course Components/ Design Principles and Goals	Content	Activities	STEM (NGSS Dimensions)
The Built Environment and Invisible Infrastructure 1) Instill ethics of Permaculture 2) Teach skills in context of whole systems thinking and design 3) Application of ecological design principles to social and cultural systems 4) Inspire through head/heart/hand integration 5) Articulate economic/jobs connection 6) Empower lifestyle changes	The Home System; cultural and social design; multi-generational living; ecovillage design; urban Permaculture and transition town movements; design for catastrophe; creative land access arrangements.	Tour of Earthhaven Ecovillage; community scale Permaculture panel; explicit emphasis on class design for multi-generational inclusion; presentation on transition initiatives; highlighting a local land management plan.	***Scientific practices:*** Observation; Discussion, ***Content:*** Science in society; Engineering; Creativity ***Crosscutting*:** Cause and effect; Systems; Stability and change
Technology and Economics 1) Instill ethics of Permaculture 2) Teach skills in context of whole systems thinking and design 3) Application of ecological design principles to social and cultural systems 4) Inspire through head/heart/hand integration 5) Articulate economic/jobs connection	Introduction to global economic dynamics. Designing a Permaculture system for economic yield. Project cost estimates and bids and budgets. Low, mid and high-tech tools and when each is appropriate.	Economics lecture. Tool show and tell; daily demonstration of ergonomic tool use on job sites; morning presentations integrating cottage industry yields into client design concepts; discussion of project costs between clients and students.	***Scientific practices*** Tool use; Discussion; Observation ***Content:*** Mathematics; Technology; Engineering; Science in society ***Crosscutting*:** Systems; Cause and effect

The Permaculture teaching and learning goals are harder to assess, as is often the case in CSL settings where participants are internally motivated and do not see the need for nor expect to be subject to external evaluation. Using the CSL strands for categories, we include here sample interview responses from those willing to take the time to discuss our teaching/learning interest. These responses suggest that participants perceived that they were taught the CSL components that were designed into the PIA program. For example, one participant quoted components that were inherent in Strand 1 – Experience excitement, interest and motivation to learn:

> The biggest challenges facing humankind have everything to do with the petroleum fiasco, natural gas in the North Sea, hydrogen sulfide, high-powered greenhouses gases being pumped out leading to climate change … the ramifications of peak oil. Permaculture ties directly into that – petroleum-based fertilizers for agriculture. We need to get the "progress is God" paradigm to die … progress being making so much money as we can no matter what the consequences. (A.M., Interview 1, 2012)

Evidence of Strand 2 – Come to generate, understand, remember and use concepts related to science; Strand 3 – Manipulate, test, explore, predict, question, observe and make sense of the world; and Strand 4 – Reflect on science as a way of knowing; on processes, concepts … on their own learning process, are evident in the following quote:

> I've learned a lot of really interesting facts about how different organisms in an ecosystem are relating to one another that are really intriguing … and different techniques for managing land systems like coppicing and installing the French drain, water uses … how I look at water and realizing it can be very very abundant if caught properly … learning how to capture and store heat. Yeah. Learning these ways to catch energy are really fascinating and really cool and pretty simple. When I look at spaces, I think this would be there and that would be here. I have a better sense of orientation in outdoor spaces. (J. Interview 2, 2012)

Another participant captured elements reflecting those of Strand 5 – Participate in science activities, learning practices with others using language and tools, in the following quote:

> I'm enjoying that sense of seeing large projects get done in one day and the power of a group of people coming together and working … and I don't even feel like I'm working that hard is the thing, because there are so many hands … and the result is inspirational and hopeful and just feels good at the end of the weekend. The intellectual side of me longs for more of the structured teaching….There's also such a beauty in just seeing something be done, and doing it and feeling it in your body, and knowing what it feels like to dig a pond, dig swales, etc. I feel like the class model is a great way to get the class needs met while also the participants' needs met. (J. Interview 2, 2012)

Evidence of Strand 6 – Think of themselves as science learners and develop an identity as someone who knows about, uses, and sometimes contributes to science, were also captured by a participant:

> I feel like in a lot of ways ecology and Permaculture are a way of thinking and a way of life and I understand that now. And I'm incrementally starting to embody that….But I'm looking forward to the journey. In terms of hard knowledge….You realize how much there is and you know none of it. I loved the hands-on hard knowledge of installing cisterns and rainwater systems … and seeing the practical applications of what to look for in plants that want to grow together and memorizing some of those. (K.C., Interview 3, 2012)

More than half of the students said their practical skills had improved and that they now felt more confident to implement Permaculture techniques on their own properties, as well as being more aware of the impact that their actions have in the larger context of global environmental issues. Students learned and reported increased retention of detailed information around various agro-ecology techniques such as mushroom gardening, annual crop production, perennial and forest crop production, poultry rearing, rainwater catchment systems, soil improvement, weed control, earthworks and hard-scaping. However, we observed through class exercises designed to test and develop their practical design sensibilities that the program had limitations. The students' design ideas illustrated insufficient understanding of the practical installation requirements and realistic yield expectations. For example, a group stated that their design would allow a landowner to "produce all her own duck feed" for a flock of 6–10 ducks from black soldier fly (BSF) larvae raised on compost material generated by her home and yard. Although the raising of BSF larvae is a proven technique for creating quality poultry feed from waste streams, the quantity would be insufficient for the landowner to produce them as the primary source of feed for a flock of 6–10 ducks. Below are photographs taken on the 2012 sites. (Figs. 6.1, 6.2, 6.3, and 6.4).

Fig. 6.1 Mushroom plugging

Fig. 6.2 Char making

Fig. 6.3 Planting

6.10 Discussion

Science education policy documents almost always propose as one rationale for the study of science that it is about preparation for lifelong learning in a changing world (e.g., NGSS Lead States 2013; Rutherford and Ahlgren 1990). The world continues to change after our compulsory education is finished and we must look to CSL to

Fig. 6.4 Group discussion

provide us with information and potentially new practices, much of which is discussed by Bell et al. (2009) in the landmark book *Learning Science in Informal Environments*. Formal education is required through approximately age 18, much less than life expectancy in terms of lifelong learning. The unique synthesis of whole systems design and ecological content taught to adults through hands-on Permaculture requires unique training and skill sets to teach (Mollison and Holmgren 1978). In this PIA program, adults have sought just this kind of learning to find out how they might manage their land, or help others to do so in a way that supports regenerative practices as a human choice. They are *feeling* the urgency of the problem and *want* to be part of the solution as the motivation outlined in Strand 1 for Informal Science Learning (CSL) by Bell et al. (2009).

We were generally successful in the goals we set for ourselves. We had also created a supportive learning atmosphere to complement the program content. Other Permaculture teachers/researchers have advocated for transforming teaching from an individual mind and effort to a collective process (Anderson et al. 2019). Throughout the entirety of our course, although we were delivering very serious content that pertained to the destiny of our species, we mostly kept it light-hearted, fun and hopeful. We provided coffee and tea in the mornings, to welcome our participants in the environment of a hearth. We cultivated a safe emotional environment for ourselves and the group in response to current news around climate change, genocide, or other suffering. The strenuous physical labor itself seemed to bring people together in an embodied way that complemented the intellectual experience.

The logistics of this kind of class in the community are complex and as in any CSL program or setting are dependent upon client satisfaction as well. We became even more aware that as teachers we have to have enough real life experience with diverse system elements to successfully micro-design each element, create detailed work plans, debrief crew leaders, and supervise a large group rapidly implementing several projects simultaneously so that they meet the standards of Permaculture sustainability. Further research in methods within this setting would be helpful.

We have learned that what we planned has enrolled people with an existing emotional connection that motivates them to seek the teaching in CSL that we have to offer. The people with whom we are working and learning have not been subsistence farmers whose lives immediately depend on this learning such as those in Kenya and rural South Africa (Arko-Achemfuor 2014; Hockin-Grant and Yasue 2017). Our middle- and working-class participants come for an optional expanded ecological vision. The students were pleased with the combination of presentation, discussion, and on-site hands-on and reality-based teaching. This process allows for learning at different levels of readiness and knowledge. Dynamic skill theory predicts that such a combination of emotional, cognitive, physical and social components will lead to learning (Fischer and Rose 1998; Mascolo and Fischer 2010; Yan and Fischer 2002). Evidence collected to date supports that there has been learning about Permaculture, but that it is not linear, as is also proposed by learning theory (Fischer and Rose 1998). We would benefit from further research in adult motivation to engage in Permaculture practices.

There are lessons to be learned from this program in terms of growing the Permaculture movement to be an adaptive part of the earth's biosystem. Some of these lessons generalize to many out-of-school science education efforts. It is possible for compulsory education to take part where renewability is part of the program – which of course we hope increases.

When planning to work with property owners, we encountered challenges that included appropriate sites, appropriate attitudes, and long-term commitment on the part of the landowners. We need to consider each of these landowner sites for their fit to our mission as well as the practical issues of good teaching sites. This is a reality that we hope will become less of an issue (or more of a choice) as Permaculture sites expand through our teaching and that of others, including those we train. Working with living systems, even in school gardens, requires experience, detailed planning and good communication/negotiation skills. Is a neighbor concerned that a beehive will lead to stings if a group (in or out-of-school) wishes to study bee cultivation? Will suburban neighbors question the looks of water catchment systems? In general, how do we approach those who look at the world differently? Having people on a project (teaching/learning) development team with good people skills is critical. This is an essential part of community networking and support as well. It is important to communicate to the greater public that there is an urgency and that there are programs in which they can participate.

For our project, long range planning is obvious to us. Landowners who make a commitment to the 10–15 years it takes to transform their properties have to be able to plan the cost out that far. There are micro-loan funds and credit unions in some very poor countries that may prove to be good models. As science educators in the U.S., there are often funds available through grants and competitions. It would be helpful to have a central clearing house for Permaculture teaching funds which would benefit the whole system (teachers, students, and landowners). A long-term plan for knowledge transfer is also necessary. Schools of education train future classroom teachers to continue the formal schooling system. Permaculture design is prepared with a 100–125 year ecological cycle in mind. Since few of us live to be 125 years old at this point in history, we are by definition designing a multi-generational system. To protect the earth, we need to address the difficult questions of commitment and property inheritance.

Many aspiring young farmers are in a difficult position because the cost of purchasing land (especially in desirable regions where demand for organic heirloom agriculture exists) has been driven to extremely high levels by development and population pressures. These young people need access to land without going into untenable levels of mortgage debt in order to become farmers, which is a notoriously fickle business proposition. Organizations like Western North Carolina Farmlink in North Carolina act as a "dating" service between landowners and potential farmers/land managers, offering contracts and paperwork around business arrangements such as leases and crop-sharing arrangements, helping to negotiate complex property issues such as land trusts, conservation easements and present-use tax valuation, and connecting farmers with the learning opportunities and mentors they need. We can suggest that schools include farming in their career awareness programs and that science departments promote the science inherent in Permaculture farming, along with their Science-Technology-Society (STS) curricula.

We are aware that adaptable teaching formats (shorter or longer time spans available to the public) increase participation. We have found four hours to be an absolute minimum time to deliver any Permaculture conceptual content and do a hands-on project grounded in that context. The two-day program was more satisfying for the teacher and students alike. The longest single class sequence we implemented in this format is 14 days, distributed over seven weekends from April through October of 2013. Ultimately, we imagine it as a basis for a type of school, training hands-on Permaculture farmers who can then start their own projects in urban and rural areas of their choosing. As we did in the second year of our program, graduates from an initial foundational program become eligible to be apprentice/students in a deeper program and to take on leadership positions in future iterations of the foundational class. Much like the common arrangement in which graduate school students teach undergraduate classes at universities, this Apprentice Crew Leader system builds in mentoring and leadership training, as well as taking some pressure off the lead teachers and allowing a lower instructor-student ratio in the foundational classes. We suggest that in any CSL program, the designers consider a model of varied length and continuity. There is a need for introductory sessions and for intensive sessions and perhaps for some in between. Offering alternatives provides for greater diversity of attendance, giving people a choice of a level of involvement. Some museums have career ladder programs where teens can be volunteers who may choose to move into apprentice explainer programs and perhaps even staff positions as time goes on. This kind of mentoring supports continuity while sharing the workload.

There is clearly much need for continued research in our program and in the field. We would like to find out more about motivation and how some people become aware and passionate about protecting future generations. We would like to know more about the relationship between school science and the science we teach. We would like to know more about how to support our work through networking. The biggest question is the ability of this program to change behaviors. Longitudinal studies are difficult and costly, but they can provide information about what happens as a result of an intervention – a program to teach (in this case) a way of treating the earth has big stakes. We would like to see more of that research happen.

As CSL teachers with a mission to educate landowners about how to manage their properties in ways that support life now and in the future, we have many challenges. Some are the challenges of all CSL: recruitment, funding, educational mission, and research/evaluation. The other issues in this chapter are specific to this setting: collaboration between landowners, teachers and students, the relationship between commitment and time, heritability of property and Permaculture principles/techniques. The adults who enrolled in our Permaculture program were current or potential landowners. They were motivated by their interest and desire to use their pieces of the earth in ways that would contribute to health for themselves, their communities, and for the future. The science that they needed was interdisciplinary—biology, chemistry, physics, and earth science often in the crosscutting dimensions proposed for science teaching in the NGSS (2013). The teachers and participants had to consider STS as they planned within urban settings that had existing policies. They had to learn about collaborative methods of installation and the economics of their choices. We have provided readers with both processes and insights within this program which is a part of what we think is critical CSL learning in one facet of our stewardship of the earth. Harnessing the theories and practices from both Permaculture field work and science education brings strength to our work and an intersectionality that we hope will add to the literature on supporting our earth sustaining mission.

References

Anderson, C. R., Binimelis, R., Pimbert, M. P., & Rivera-Ferre, M. G. (2019). Introduction to the symposium on critical adult education in food movements: Learning for transformation in and beyond food movements—the why, where, how and the what next? *Agriculture and Human Values: Journal of the Agriculture, Food, and Human Values Society, 36*(3), 521–529.

Arko-Achemfuor, A. (2014). Teaching permaculture to ensure food security in rural South Africa: The case study of tiger kloof. *Journal of Human Ecology, 47*(3), 251–255.

Bell, P., Lewenstein, B., Shouse, A. W., & Feder, M. A. (2009). *Learning science in informal environments, people, places, and pursuits*. Washington, DC: National Academies Press.

Darimont, C. T., Fox, C. H., Bryan, H. M., & Reimchen, T. F. (2015). The unique ecology of human predators. *Science, 349*(6250), 858–860.

Fischer, W., & Rose, S. (1998). Growth cycles of brain and mind. *Educational Leadership, 56*(3), 56–60.

Hockin-Grant, K. J., & Yasue, M. (2017). The effectiveness of a permaculture education project in Butula, Kenya. *International Journal of Agricultural Sustainability, 15*(4), 432–444.

Holmgren, D. (2002). *Permaculture: principles & pathways beyond sustainability*. Hepburn, Vic.: Holmgren Design Services.
Immordino-Yang, M., & Damasio, A. (2007). We feel, therefore we learn: The relevance of affective and social neuroscience to education. *Mind, Brain, and Education, 1*(1), 3–10.
Katz, P. (2017). Formerly ISE: Preparation for continual science learning. In P. Patrick (Ed.), *Preparing informal science teachers* (pp. 13–38). New York: Springer.
Lebo, N. III. (2011). *Towards ecological literacy: A Permaculture approach for junior secondary science.* http://transitionvoice.com/2011/12/towards-ecological-literacy-a-Permaculture-approach-for-junior-secondary-science/
Marsh, C., Allison, P., & Friedman, Z. (2014). *Permaculture Playbook*, available by request through zevkudzu@gmail.com
Mascolo, M., & Fischer, K. W. (2010). The dynamic development of thinking, feeling, and acting over the lifespan. In W. F. Overton (Ed.), *The handbook of life-span development* (pp. 149–193). Hoboken: Wiley.
Mobus, G. E. (2018). Teaching systems thinking to general education students. *Ecological Modelling, 373*(C), 13–21. https://doi.org/10.1016/j.ecolmodel.2018.01.013.
Mollison, B. C., & Holmgren, D. (1978). *Permaculture one: A perennial agriculture for human settlements*. Melbourne: Trasworld Publishers.
NGSS Lead States. (2013). *Next generation science standards: For States, by States*. Washington, DC: The National Academies Press.
Palumbi, S. R. (2001). Humans as the world's greatest evolutionary force. *Science, 293*(5536), 1786–1790.
Permaculture Worldwide Networks. http://Permacultureglobal.org/projects?page=3&type=Urban
Permaculture Research Institute. (2020). http://Permaculturenews.org
Praetorius, P. (2006). Teaching through permaculture experience. *Connect Magazine*, Nov–Dec., 5–9
Rios, M. (2010). How to add zest to your sustainability education program. *Communities*. Summer 47–51
Ritchie, H. (2018). Urbanization. *OurWorldInData.org*. Retrieved from: https://ourworldindata.org/urbanization 29 Nov 2020
Rutherford, J. F., & Ahlgren, A. (1990). *Science for all Americans*. New York: Oxford University Press.
Stamets, P. (2005). *Mycelium running*. Berkeley: Ten Speed Press.
The Royal Society. (1985). *The public understanding of science*. London: Author.
United Nations Department of Economic and Social Affairs. (1987). *General assembly plenary meeting*. http://www.un.org/documents/ga/res/42/ares42-187.htm
Vitousek, P. M., D'Antonio, C. M., Loope, L. L., Rejmanek, M., & Westbrooks, R. (1997). Introduced species: A significant component of human-caused global change. *New Zealand Journal of Ecology, 21*(1), 1–16.
Wells, P. (2011). The transition initiative as a grass-roots environmental movement: History, present realities and future predictions. *Interdisciplinary Environmental Review, 12*, 372–386.
Yan, Z., & Fischer, K. (2002). Always under construction. *Human Development, 45*(3), 141–160.

Chapter 7
Learning to Become "Good Food" Educators: Practices and Program Development of an Urban Agriculture Education Organization

Christopher D. Murakami and Heather Gillich

Abstract In 2009, The Urban Agriculture Education Organization (UAEO) was founded by a group of environmental and food system activists to provide food and garden-based education programming. This study focuses on the organizational learning and change within their community of practice from 2008–2014. Narrative interviews were conducted with UAEO founders in 2011 and education program employees in 2014. The chapter presents the "healthy soil ecosystem system" model that is an extended metaphor to describe the practices of the education organization and describes the interrelated practices of (1) participating in the community ecosystem, (2) balancing fertility, (3) improving structure, and (4) enhancing biotic activity. Each of these four themes is described and situated within the story of UAEO's educational program development and helps describe what it means to be a practitioner of urban agriculture education for the organization over time. These practice areas help highlight what was valuable for the organization and our analysis helps describe how organizations consider core practices and continue to reflect and improve programs over time. The implications for urban agriculture and science, technology, engineering and mathematics (STEM) education practitioners and research are discussed.

Keywords Non-profit organizations · Organizational change · Garden-based education

C. D. Murakami (✉)
Chatham University, Pittsburgh, PA, USA
e-mail: c.murakami@chatham.edu

H. Gillich
Healthy Living Initiative, City of Minneapolis, Minneapolis, MN, USA

I. DeCoito et al. (eds.), *Teaching and Learning in Urban Agricultural Community Contexts*, Urban Agriculture,
https://doi.org/10.1007/978-3-030-72888-5_7

7.1 Introduction to the Urban Agriculture Education Organization

The Urban Agriculture Education Organization (UAEO) is a 501(c)3 educational non-profit organization that was founded in 2009 in Midwestern University City, USA (pseudonym), by a group of environmental and food system activists that brought the central Midwestern state community together around the notion that "*food is good.*" This organizational tagline represents the idea that working towards social and environmental justice is made less polarizing and more engaging for learners of all ages when explored through good, delicious food. Importantly, good food was never fully defined by the organization and provided a unifying and intentionally apolitical approach to food education. Chris Murakami (the first author) started studying this organization in 2010 and was part of ongoing embedded qualitative research from 2010–2016 to explore the practices of this urban, sustainable agriculture education organization and its associated programs. This chapter focuses on the period between 2008–2014 during which the organization grew substantially in the depth and breadth of learning experiences that were offered to diverse community members of all ages. Heather Gillich (second author) joined the organization in 2012 and facilitated some of the programming expansion. In 2016, UAEO had eight full-time and three part-time employees who helped support urban agriculture experiences for youth on their 1.5 acre urban farm and outdoor classroom projects based in nine elementary schools, one high school, and two non-profit agencies. During the 2014 growing season, UAEO reached 866 Pre K-12 learners during on-site urban farm experiences like field trips and workshops and 1336 students in off-site outdoor classrooms. As we reflect on this period of time, now more than 4 years after the initial research was conducted, we acknowledge that with major social movements such as Me Too, Black Lives Matter, and the COVID-19 pandemic, we think differently about the most useful frameworks for understanding educational systems and organizations, especially those dedicated to food education. With this caveat, we present this chapter that explored the programmatic features that helped UAEO make progress towards organizational goals of supporting active participation in the responsible management of human and other natural resources while supporting K-12 science, technology, engineering and mathematics (STEM) learning.

Etienne Wenger (1998) describes learning as a process of becoming a member of a community of practice. To start this case study, exploring the UAEO community of practice and its education programming from 2008–2014, my co-author and I briefly share our initial experiences. This introduction shares some of who we were as authors engaged in this research and also provides a glimpse into UAEO, an ever-changing urban agriculture educational, non-profit.

Chris: (July, 2010) After finally settling into my house just north of downtown Midwestern University Town, USA, a city of 130,000 and home to the Midwestern Town University where I would start a PhD program in Science Education in the fall, I went on a walk to explore the neighborhood. It was a Thursday evening, towards the end of July, and my wife and I came upon a group of urban farmers working in what they called their "demonstration garden." We are trying to, "demonstrate the viability of urban agriculture," one of the three bearded, tall, presumably white, twenty-something year olds stated, a line he had undoubtedly rehearsed and regurgitated numerous times. The demonstration garden sits on an abandoned corner lot owned by a generous local student housing real estate group. There were bicycle wheels attached with rusted baling wire to a fence surrounding what was the most productive and vibrant vegetable garden I had ever seen at the time. I was given an informal tour by Greg (pseudonym), one of the three founders, he was eating purple podded pole beans raw from the vines, tossed me one and said, "this is good, you should eat it." I told them my interests in garden education, sustainable agriculture, and my plans to study both while pursuing my PhD. In a later interview, Greg recalled it, "as the clouds parting," when I stumbled upon the garden and UAEO. I helped fill a void in their educational programming which their nascent non-profit was based upon, despite a lack of formal training or experience by any of the founders or "staff." (In quotations because at the time there were not any paid positions, with all of the members of UAEO living in "poverty by choice" in a large Victorian era house – also the UAEO headquarters). As a volunteer, I started showing up, asking questions, pulling weeds, facilitating reflection, and offering direction. I started gathering data, informally at first, trying to understand what fueled this growing movement and passion for urban agriculture education. Later, I realized that I had "gone native" and was part of the UAEO community, but still interested in making meaningful theoretical and practical contributions through research for the organization and the broader movement dedicated to connecting people to food.

This chapter is a product of research and practice in urban agriculture and STEM education that started for me on that sticky Thursday evening in July and paralleled my PhD training in science education. While the organization focused on providing learning experiences in urban agriculture to support STEM and environmental learning, the analysis in this study focuses more on the organizational learning and change within the community of practice to examine the underlying practices and approaches that shaped educational programming from 2008–2014. The direction of the organization was shaped in part by the dedication and passion from Heather (second author), UAEO's Education Director from 2012–2017, who came to UAEO in March of 2012 as an AmeriCorps Volunteers in Service to America (VISTA) member. One of the important organizational changes for UAEO was receiving an Americorps VISTA grant in 2012 that partially funded three full-time positions to help the organization achieve their mission. Heather was hired to serve as the "education VISTA," a job that entailed creating the educational programming capacity for UAEO. There was not a formal predecessor for Heather's position other than the founders who each identified as educators in some form or another and Chris' contributions on a volunteer and piecemeal basis through several small grants. The initial educational leadership and vision of the full-time employees like Greg were providing urban agriculture education experiences but still learning how to build strong partnerships and educational programs. Below, Heather shares a story of her initial impressions of the organization and her new job that she moved across the country to start.

Heather: (March, 2012) I was excited. I packed my truck with my belongings and my dog, left the comfort of my parent's home, and embarked on my journey halfway across America. After a grueling full day of driving, I was less excited. I woke up my dog and we began to explore our new, if temporary, home. I was pleasantly surprised to find a delicious restaurant for brunch, a coffee shop full of engaging conversation, and more than a few community gardens beginning their entrance into spring. I was getting that feeling of excitement back. On my first day as an Americorps VISTA, I went to my new "office" only to be greeted by a large, smelly German Shepherd. Where am I? I felt tricked. The people I had talked to on the phone were well-spoken, professional people! Nonetheless, my office was in the home of one of the founders, and that large German Shepherd was to be one of my office mates. My other office mates were those well-spoken, professional people I had talked to, though I could tell that professionalism was only one of the many hats they wore. Soon I joined them in the dirt and found that I was exploring myself as much as this community. My initial assignment was to bring 100 students to the Urban Farm. What students? From where? No one seemed to know. It turned out that I needed to start this program from scratch, with only the relationships built by the organization's founders to guide me. Whenever I talked to Midwestern University Town community members about getting involved with UAEO's "good food education," I was pleasantly surprised by how many of them shared my excitement. Those office mates of mine had done great work building relationships, and I began to try very hard to maintain them. When I came to UAEO, I never expected to want to stay. Given the opportunity, I was thrilled that I would continue to build on those local relationships, and that I would be able to bring some of my ideas to reality. Our educational programs in 2014 served nearly 6000 participants, and I haven't lost that feeling of excitement. Though my days at work are much different than they were in that first year, and include increasingly more paperwork, I feel grateful to do this work with such an amazing group of people.

7.2 Urban Agriculture and Garden Education

These two brief stories of coming to the UAEO community captures two time points in the organizations evolution. It is important here to bring attention to the early idealism and vision of the organization that surely motivated many other readers and food activists, but also the harsh reality of trying to build the organizational and community infrastructure necessary to make progress towards legitimate social change. As urban agriculture educators and researchers, we intend to share contributions that elaborate how the organization came to make meaning of its practices in a particular situated context. Part of that situated context is the political and economic conditions that engendered the creation and modification of UAEO as well as important theoretical and practical contributions of scholars, practitioners and good food education activists. Here, we briefly review literature that frames our inquiry, while also shedding light on some of the important contextual factors during that time.

In this book is a compilation of other pragmatic food system activists and organizations like UAEO who are engaged in the challenging but fruitful work of using food education to support STEM learning, part of a growing movement of what Mark Winne (2010) refers to as *Food Rebels*. Like many other food rebels, UAEO's founders were concerned about the status of global food systems and their dependence on an infrastructure that runs on cheap fossil fuels, ecological degradation, and exploitation of human labor. Despite adequate food productivity, there is an alarming rate of food insecurity. Before the COVID-19 pandemic and widespread economic disturbances, the United States Department of Agriculture Economic Research Service estimated that in 2019 13.7 million households (10.5%) experienced some level of food insecurity (Coleman-Jensen et al. 2020). These issues of food access are rooted in poverty which has also worsened of late with growing disparities between the wealthy and poor citizens. Access to higher education and specifically STEM education, is one way for students from under-resourced backgrounds to eventually attain the training and desire to pursue careers with higher earning potentials (Olson and Labov 2014).

With a growing emphasis on STEM education reform, many garden educators are well-positioned to engage school districts, funding agencies, and students in the production and preparation of foods, while possibly recruiting and retaining the next generation of STEM workers. Like many other urban agriculture organizations, UAEO was formed during a time of economic recession in the US. Early on within the UAEO community of practice, they were dealing with the reality that building a non-profit organization required significant economic sacrifice and hardship. Many of the employees in 2016 were motivated by the same visions of sustainability education and good food education that in some ways might be in contention with the rhetoric and broader philosophical implications of STEM education reform. However, it is important to note that STEM education reform has the possibility of narrowing the vision of science education (Zeidler 2016). Scholars like Zeidler question the narrowing effect of STEM reforms that are heavily focused on job preparation. They argue that without the arts and humanities, science education might fail to provide learners with the tools necessary to adequately participate in the decision-making or social transformation that might be requisite for global and domestic progress towards social and environmental justice. More functionalist or neoliberal perspectives on science education focused on career preparation might insufficiently provide opportunities for learners to think critically about the broader oppressive structures within societies and communities that need to be transformed to make progress towards social justice and sustainability through advances in, or applications of, STEM.

Weaver-Hightower (2011) poignantly articulated reasons educational researchers should explore the role of food in educational institutions because it sheds light on issues of social and environmental justice. This chapter focuses on UAEO, an organization that is devoted to "good food education" in order to be more intentional about the ways pre-K-12 students learn about growing food. Previous research from Blair (2009) provided a systematic review of empirical studies of garden-based education and reported initiatives can improve environmental attitudes, science achievement, school engagement, and willingness to try fruits and vegetables. While there is the ongoing need to know more about program features that lead to knowledge gains and behavior changes, in this chapter we focused instead on the development of an organization that formed to provide these experiences in garden-based urban agriculture STEM education.

7.3 Theoretical Lens

To guide this inquiry, we used a communities of practice theoretical lens to better understand what it means to become a practitioner of urban agriculture education. For Wenger (1998) and Lave and Wenger (1991), this notion of a community of practice provides a unit of analysis to reorient studies of learning that extend beyond the realm of schooling. For example, in our study, we are examining the broader organizational learning and patterns by trying to make sense of the practices, patterns and changes of the educational programs over time. This undoubtedly has an impact on the nature of learning that would unfold in classroom or garden-based settings, but the K-12 learner experience was not the focus of our analysis. Many scholars in science education have been influenced by communities of practice and other sociocultural learning theories to explain how processes of identity formation happens and how identity trajectories are authored in informal environments such as after school programs (Tan et al. 2013). In this study, we are applying this communities of practice lens to look at the organization itself, specifically its education department, to offer an opportunity for self-reflection and growth. Throughout the research process, insights were shared with leaders in the organization to help them serve the needs of the organization as part of iterative planning. Here, we share the process of reflection with a broader audience of urban agriculture education practitioners with the hope that it resonates with their experiences but also sheds light on some approaches to organizational change for urban agriculture and STEM education.

Communities of practice share values that are always changing and influenced by decisions and dilemmas within the group (Wenger et al. 2002). Examining these organizational values and priorities over time can help clarify what it means to practice urban agriculture education. Overall, this study addresses the following research questions: What does it mean to become a practitioner of urban agriculture STEM education for UAEO? How did UAEO's educational programming change over time?

7.4 Methods

This research is guided by case study methods (Stake 1995) and iterative qualitative methods (Coffey and Atkinson 1996) and focuses on UAEO's education department and programming as it has changed during the period from 2008–2014. UAEO has a multipart mission and diverse educational programming for learners of all ages, but for the purposes of this case study, our analysis focuses only on the Pre-K to 12 educational programs. This programming has changed over time and includes on-farm learning experiences, in-school classroom presentations, outdoor classroom collaborations, and afterschool programs. This research occurred in two main phases. During the first phase, the first author conducted narrative interviews with the three founders, Greg, Jeff, and Steve (pseudonyms). These interviews told the story of the creation of UAEO from the three-year period between 2008–2011. The stories and experiences embedded in the founders' narratives were used to help develop a model for the underlying educational practices in conjunction with the early educational programming of this nascent urban agriculture education nonprofit. Interviews were transcribed and open-coded using NVivo 10 qualitative analysis software. In searching for the founders' underlying practices and approach, patterns emerged that resonated with different approaches to education or underlying purposes of education. These orientations were later used to develop the "healthy soil ecosystem model" to describe the practices of the urban agriculture educators. Both Greg and Steve were consulted during follow-up member-checking interviews (Cho and Trent 2006). Jeff was not available for member-checking. Our analysis was meant to highlight the values that the founders of the organization held related to urban agriculture education, and serve as a historical foundation for UAEO to build upon.

During the period from 2011 to 2014, UAEO underwent substantial expansion and reorganization of their education department. Heather's story of coming to UAEO as a newcomer to the community of practice is a snapshot of the organization and education department that Heather began cultivating as the first formalized education director. In 2014, the preliminary findings were developed into a framework and then validated by employees with roles in the education department organization (Kelly (pseudonym) – Urban Farm Manager, Mary (pseudonym) – Outdoor

Classroom Manager, Ross (pseudonym) – Americorps Education, and Paul (pseudonym) – UAEO General Manager). The authors of this study worked together to examine some of the organization's practices while also looking for a way to reify what it means to practice urban agriculture STEM education with UAEO. This research/practice collaboration was intended to benefit the education department, the organization as a whole, as well as urban agriculture education teachers and researchers. Throughout this chapter we layer stories (Tamboukou 2008) from the founders of the organization in contrast with programs that represent significant expansions within the education department. We sought to collapse a rich history and innumerable experiences engaging youth in urban agriculture education by providing representative ethnographic vignettes (Maanen 1988) that help explain key findings and represent salient tensions that we hope resonate with urban agriculture educators in diverse contexts. To conclude our chapter, we discuss some implications for practice.

7.5 Findings

The first phase of research based on the narrative interviews of the founding members of UAEO helped describe the multidimensional practices of becoming an urban agriculture education organization. In general, becoming a practitioner of urban agriculture STEM education means being a manager of "healthy soil ecosystems." This healthy soil ecosystem system model collapses the forms and practices of the education organization embedded in the early motivations and work of UAEO. Figure 7.1 below illustrates four interrelated themes: (1) participating in the community ecosystem, (2) balancing fertility, (3) improving structure, and (4) enhancing biotic activity.

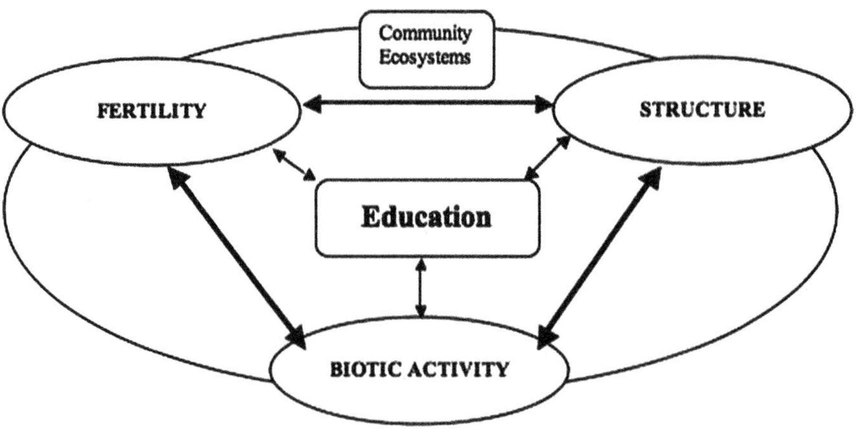

Fig. 7.1 Healthy soil ecosystem model

Participating in the community ecosystem represents the awareness that urban agriculture educators at UAEO had for the historical, political, economic, cultural, and ecological features in the situated context. The metaphorical ecosystem includes the living and non-living factors in a mutually constitutive relationship, and this practice of engaging and interacting requires ongoing relationship-building and decision-making to figure out the broader role of the organization and how it ultimately utilizes financial resources and volunteer and staff engagement to serve the broader community.

Balancing fertility is the work of simultaneously encouraging engagement, excitement, and creativity while working to make progress towards a more sustainable food system. Within the soil ecosystem metaphor, this means amending mineral resources and nutrients to support metabolic function and growth. This theme represents the physical resources or human motivation necessary to fuel the work of urban agriculture education. This theme embeds educational philosophies that support the liberal arts and humanities because of their ability to encourage critical and systems thinking, informed decision-making and supporting learners as they develop identities as active democratic participants and stewards of natural and human resources. With a lens towards justice, it is important to assure that access to these resources is equitable and inclusive and does not perpetuate patterns of exclusion, underrepresentation, or deficit-based narratives.

Improving structure in soil science means managing the arrangement of particles and organisms in the soil ecosystem that help enhance the movement of water, air, and nutrients. For an educational organization, this means aligning people's interest and skills within defined roles to better achieve organizational or program goals. One of the many purposes of education is to help provide structures within learning environments to support the socialization of students, to later help them find a position within society.

Finally, enhancing biologic activity represents increasing the overall productive capacity of a soil ecosystem, which is maximized when there is a proper balance of fertility and appropriate soil structure. This theme represents some of the educational outputs or other products that are enhanced when structure and fertility are properly managed. We intend this theme to elicit reflection in terms of what outputs are valuable to help think about program and educational evaluation. This biologic activity also has the capacity to act through positive feedback loops to affect each of the other themes.

We believe that each of these themes are integrated and altogether represent the practices of urban agriculture educators who are hoping to scaffold learning experiences to encourage active participation in the transformation of the food system. Next, we present vignettes from two different time periods in the organization (2008–2011) and (2012–2014) to illustrate each theme and explain how they represent organizational change and maturity. Throughout the next section we have purposefully contrasted examples of the early education programming with later programming to allow the reader to notice the ways that the organization has evolved over time and cultivated a strong identity as urban agriculture educators for the city and region.

7.6 Participating in the Community Ecosystem

Participating in the community ecosystem means taking advantage of the resources and connections made available in a situated context. Through a series of mutual friends, a parent at the school expressed interest in getting help from UAEO to start offering lessons to children at a small, Private College Lab School training program. In 2009, UAEO was able to negotiate an agreement to provide regular lessons in exchange for access to an underutilized greenhouse (in the background) to grow plants for their 1.3 acre urban farm and 1/12 acre demonstration garden. Both of these sites were owned by a local landlord who rents to college students in the city. During these early phases of education programming, the organization was extending its reach out into the community and looking for ways to gain access to resources and share their passion and novice experiences of growing and accessing foods. Greg explained, "we were throwing things against the wall and seeing what sticks," recalling the early attempts to propose projects and programs, build connections with groups in the community and cultivate mutually beneficial relationships. Things stuck when there was shared investment, new funding opportunities, or new partners to achieve somewhat improvised goals or programs. There were countless examples of UAEO's grassroots and community-driven development, but this example of early programming at a College Lab School represents this idea of filling ecological niches and forming symbiotic relationships. Over time, however, these types of partnerships established through informal agreements and community connections were modified because of the tendency for the community partners to expect UAEO to provide these educational services for free.

The passionate teacher/parent/volunteer model of urban agriculture and garden education is a powerful and often necessary starting point, but in most cases can fall short when trying to make progress towards larger scale, longer lasting change. In thinking about ways that the role of UAEO has changed over time when providing learning experiences to K-12 students, the Outdoor Classroom Project coordinator, Mary, realized that she had to think beyond just volunteering her passion and energy to provide lessons for teachers or partnering schools. She started her full-time, paid position with UAEO in 2014 and explained that she had to transform her thinking and general approach to urban agriculture education. After providing hundreds of lessons and talking to countless others in the community, UAEO began to get an idea they could do more than just provide lessons. Mary explained, "I'm not providing a service anymore – I am building a program," which requires a reorientation that goes beyond offering one-time lessons to think about packages of more developed curriculum. This meant learning how to develop program goals that resonated with receptive audiences while holding true to UAEO's mission.

In 2016, the Outdoor Classroom Program had changed UAEO's relationship with the Midwestern University Town Public School District by extending UAEO's reach beyond the Urban Farm and onto school campuses. Rather than providing lessons in schools, UAEO works with Outdoor Classroom Committees made up of representatives at each school who develop autonomy and facilitate their own programs. New partnerships were formed with Slow Food Katy Trail, another local nonprofit, to administer their Harvest of the Month Program (a program that provides on-site cooking and farmer interactions to learners). This package, in conjunction with the support of UAEO in the Outdoor Classroom Committees, was meant to meet the needs of teachers by helping to meet required learning outcomes through longer-term learning experiences. UAEO is still learning how to navigate and negotiate relationships with schools and teachers to figure out how to best collaborate to serve common interests. Both Mary and Ross have training and background in education and have been able to support curriculum development that aligned with the values that partnering schools hold, and meet partner needs. Furthermore, requiring the establishment of committees at each of the schools helps ensure local ownership and management of the on-school landscapes and supporting program sustainability that can work to enhance the individual, passionate teacher model. This evolution in UAEO education program development shows the continued need to seek out this type of symbiotic relationships with the broader community ecosystem to help the program expand and meet the needs of diverse school contexts.

7.7 Balancing Fertility

In 2008, Greg, Steve, and Jeff were recent graduates from the Midwestern Town University and began dreaming up and enacting positive change for the world that started "in [their] own back yard." In all of the founder's stories of what drove the creation of UAEO, they explained an underlying motivation to make progress towards a more sustainable food system, by taking pragmatic, local steps. With a small grant for materials, they were able to buy bike trailers, t-posts, and seeds. They started collecting food scraps on bicycles to make compost to add important nutrients to the soil. The founders looked to examples from history and knowledge of science to see the potential for urban agriculture to address some issues like climate change, a global food system dependent on fossil fuel, and a generation disconnected from the soil and their food. All of the interviews mentioned inspiration from, "the special period in Cuba," when Cubans learned to practice urban agriculture to survive following the broader food system failure as a result of isolationist policies and the fall of the Soviet Union. Before UAEO existed formally, the founders offered a service-learning course for college students that primarily focused on making compost and starting urban gardens. "At first we were focusing on sustainable development for downtown Midwestern University Town," Steve recalled during the interview. In 2011, after partnering with the Midwestern University Town Housing Authority they received a sizable grant to install fruit trees in downtown housing authority sites and to offer an afterschool gardening program. UAEO received one of its first grants from the Housing Authority that provided fertility in the form of financial capital to pay wages, while also making progress towards expanding opportunities for urban agriculture and sustainable development in the form of edible landscaping and educational programming.

Children from families who lived in Midwestern University Town Housing Authority dwellings attended an afterschool program, *Growing Ahead (pseudonym)*, and participated in weekly gardening experiences. The first author helped engage up to 40 children (1st-8th grade) in the process of designing, building, maintaining, and harvesting from the garden during weekly lessons in 2011. This first extended educational program was eventually discontinued when funding for the programming ended the following year. While the urban agriculture educators had plenty of internal motivation, the program did not lead to sufficient capacity building or buy-in with the Growing Ahead partners. There were challenges with the longevity of the program without allocated UAEO staff time to the program, and limited time with the staff and teachers in the after school program. At the start of 2012, this afterschool gardening program no longer had grant funding to support staff time and the model of off-site programming (without funding) meant the program did not continue. The organization then focused primarily on educational programming on-site at the Urban Farm. This is a historical example that cautions educational program expansion into a region without attending to properly balancing fertility or community buy-in and longer term commitments to cover the costs of staff time and professional development.

In 2013, UAEO had a unique opportunity to partner with the Midwestern University, This is Nature NSF-GK-12 program, the Midwestern University Town's Parks and Recreation Department, and the Midwestern University Town Public School District in a consortium tasked with creating a program that encouraged learners to take STEM learning from the classroom to the community. The program was called BGREEN for University Town and it provided training for teachers and undergraduate research assistants in the classroom. This version of on-school program development for UAEO provided an important contrast to the failed Growing Ahead garden program in 2012.

One of the partnering classrooms in a local elementary school helped UAEO address a perennial issue for urban and school gardens. The cold winters when the garden is dormant and warm summers with prolonged dry periods and clay soils make gardening challenging. These elementary school learners worked for a school year to research, test, and design a hugelkutlur garden bed. Hugelkultur is a process of mounding logs, limbs, leaves, grass clippings, topsoil, and compost to create a structure good for growing both annual and perennial plants. Students who worked on this project shared their results at a city-wide research symposium on campus at the Midwestern Town University. For their presentation poster, they explained that the project aimed to "find out if a hugelkultur could support a garden without irrigation or fertilization." Learners involved in the hugelkultur project developed their own method for investigating and installing hugelkulturs. They first constructed and tested small models by monitoring temperature, moisture, and plant growth. After deciding on structure, materials, and orientation, they visited the Urban Farm to plot out and install a hugelkultur in our Urban Farm Outdoor Classroom. After a successful install, they placed two more hugelkulturs at their school.

The hugelkultur program and the BGREEN for University Town collaboration represent a more intentional approach to balancing fertility. This learning experience helped engage elementary learners in STEM to address a community issue, and actively participating in managing resources on their school site. Because of the teacher and student buy-in and developed capacity and expertise, learners were able to enact urban agriculture education through STEM practices and increased ownership.

7.8 Improving Structure

Becoming an urban agriculture educator at UAEO meant finding an effective arrangement of staff that could help divide up the complex challenges of growth and development of a non-profit organization. The initial application for non-profit status was filed in 2009 and officially granted in the spring of 2010. At the time the board members were all volunteers, but they were also the full-time members of the organization. Steve explained in an interview the changing structure of the governing board:

> We were a very active board, like the board did all of the work. All of the board were volunteers – and making it happen. And no one was getting paid. Well, then as we, we wanted to make sure that people got paid. So we wanted to add that to it, so, we worked on that a lot during 09 and to figure out [...] we were working on the structure a lot and we wanted to have some coordinators [...] they would get paid to do the day-to-day management and the board would stop being a hands-on active board and be more of like a consent board. (Steve, Interview, October, 2011)

This excerpt shows again the dedication and sacrifice of the initial board members of the organization, but also represents a transition away from a "hands-on active board" where everyone who was part of the organization was involved in all of the decisions that were made. Greg explained this initial phase of the organization was exhausting and was causing fatigue for a number of board members:

> We met all that time – UAEO did – the original people. All the time we would just talk for hours and hours and hours. We were never off. We were either working on UAEO or talking about UAEO or sleeping. Like, that was it. (Greg, Interview, October, 2011)

Further, Greg elaborated that the restructuring of the board to be more hands-off and the development of coordinator positions that would be paid (starting in 2011) meant that people could naturally assume different but complimentary roles within the organization that suited interests and skills sets. He reminisced, "umm, when we first started, everyone just did everything." Greg went on to explain the roles that he felt each of the founders played, and then explained how each addition to the organization has added new dimensions and capabilities to help the organization evolve. This organizational progression has continued to stretch vertically, though like many other non-profits, UAEO began with a horizontal structure that eventually led to workers being spread thinly across numerous projects (e.g., landscaping crew member and educational programs director).

The structure of UAEO's programming and education department continued to evolve to meet the needs of the expanding community of learners. As the shift from "services" to "programs" occurred, it became important to clearly outline staff roles in these programs. Like many passion-fueled jobs, everyone in the organization still wears many hats. Staff days may include manual labor with volunteers, providing lessons to learners, organizing meetings with committee members, and giving presentations to funders. Despite the many roles that UAEO employees all play, the hierarchy of decision-making is better outlined in their organizational structure and culture.

In 2014, the Outdoor Classroom Project, initially funded and managed by a separate community group, became an extension of UAEO's programming. This restructuring and aggregation represented a new set of challenges and opportunities. To support this type of program expansion, Heather felt the increasing hierarchy within the organization and the department became necessary. However, building beneficial arrangements with the community collaborators at school sites meant that the roles of these Outdoor Classroom Committees needed to be intentionally designed to help them clearly define their own roles and the roles of learners in their Outdoor Classroom spaces. Mary was hired in 2014 to develop a new identity for the outdoor classroom project under the wing of UAEO's education department, while also bringing new energy and skills to UAEO. Mary explained that this transition was difficult for her, "I [am] now part of an organization that has various departments … with an organizational flow chart!" (Mary, Interview, April 2015). However, she has come to see this reorganization and part of a new structure that can help expand learning opportunities for students.

7.9 Enhancing Biologic Activity

Balancing fertility and improving structure help support biotic activity and productivity of a soil ecosystem. This final dimension of the healthy soil ecosystem model is meant to represent the products or outputs of urban agriculture education and progress towards organizational goals. Like alluded to in other stories of early education programming in UAEO, the focus was initially on providing lessons to a number of students. Tracking the number of learners reached was a simple and sufficient way to report progress for some of the small grants that helped establish educational programming. For example, for a Youth Educator grant from the USDA – Sustainable Agriculture Research and Education, the goal was simply to provide a series of three lessons to 10 different classrooms. As an organization, UAEO also monitored progress in terms of pounds of vegetables produced, the value of retail sales, numbers of volunteer hours served, etc. These measures provide a snap shot of the reach and programs that the organization had, and also represents some of the value that the organization brings to the community ecosystem.

However, becoming an urban agriculture educator for UAEO has meant more than just increasing the number, frequency, and duration of experiences connecting learners to agriculture and supporting STEM learning. As the structure of the organization has allowed for program expansion with strategic partnerships, and fertility has been increased with associated funding and grant opportunities, the productive capacity of the organization has successfully recruited even more participation and engagement in a positive feedback loop. The good food education opportunities are growing. This increased participation and engagement and associated identity formation is what can allow continued growth and wider reaching social change through urban agriculture education. However, quantifying or describing this enhanced biologic activity is a perpetual challenge, especially if it were to venture into more transformative attempts to address systemic issues within the food system.

In 2016, with a team of strategic partners and restructured education programs and departments, UAEO was well-positioned to think more holistically about how to monitor, evaluate, and assess the impact of their educational programming. For example, because of their funding and partnership with the regional United Way in 2012, they have the capacity to gather demographic and other data on the K-12 learners who have participated in a variety of educational programming experiences over time. By improving internal data management systems and developing new tools to monitor and document learners' experiences, there will be opportunities to explore new research questions or share case studies that can help evaluate the effectiveness of programming as it relates to metrics like student attitudes, engagement, standardized test scores, graduation rates, or attendance. While the validity or merit of these metrics might be debatable, administrative decision makers from partnering school districts will value these data, which may lead to increased fertility and excitement. This dimension of becoming an urban agriculture STEM educator is critical because it helps make the impact of the learning experiences palpable for other collaborators.

7.10 Summary

So far, we have provided a description of four dimensions that explain the practices of urban agriculture and STEM education drawing from historical examples of the UAEO that represent the maturity and ongoing evolution of UAEO's education department. These dimensions are participating in the community ecosystem, balancing fertility, and improving structure together help enhance biotic activity. These interdependent practices have changed over time as contextual factors and personnel have changed. In the process, the organization's role and the broader connections it has to the local and global community has also changed over time.

Overall, our aim of this chapter was to develop a preliminary model for understanding the complex work of urban agriculture educators. Table 7.1 summarizes some of the evolving practices of the organization and some of the examples that were drawn upon in the chapter. While this model was derived from the specific experiences of participants within UAEO during a specific time period, we hope the reader can use the story of UAEO's educational program evaluation to better

Table 7.1 Health soil ecosystem themes and evolving UAEO practices

Theme	Evolving practices	Examples
Participating in community ecosystems	-Forming partnerships -Assuring mutual benefit and investment -Determining the value of the services of the organization	-Grant-funded programs -Partnerships with shared goals and investment
Balancing fertility	-Finding alignment with organizational goals and community partners -Enacting strategies for organizational and learner ownership and investment	-Growing ahead -This is Nature
Improving structure	-Establishing self-defined organizational roles and responsibilities	-Organizational structure and specialization established over time
Enhancing biotic activity	-Identifying key outputs, activities -Enacting evaluation strategies aligned with organizational/ stakeholder priorities and programs -Continuously improving programs and functions based on findings	-Transition from grant-based metrics to organizationally aligned evaluation systems

understand their own work as practitioners. We intended this model to highlight the importance of organizational, educational, and program development philosophies and principles to form mutually beneficial partnerships. For the participants in this study, this reorientation was transformative and encouraged organizational capacity building that they continue to build upon. This capacity building was greatly facilitated by developing transparent roles for employees within the organization and partnering organizations. While these roles might be constantly negotiated and ideally changing, urban agriculture educators should have a clear understanding of the role of facilitating and supporting active participation for learners and community partners. UAEO's expanded capacity was greatly enhanced by a variety of tools that help achieve each of the four practices associated with becoming an urban agriculture educator. Some of these tools are simple, such as sign-in sheets, release forms, and surveys, but taken together can help document, evaluate, and represent the organization's biotic activity.

7.11 Implications for Practice

This chapter described some of the key practices that proved valuable to the organization during its development and growth of educational programming. Moving forward, it may be fruitful for UAEO to ask some of the following guiding questions and develop appropriate metrics and assessment tools to guide continuous improvement:

- Are partnerships with other organizations mutually beneficial and equitable?
    - Ideas for Practice: Formal reflections, conversations, or interviews with partners and openness to call-in any concerns.
- Are employees motivated by the approach to teaching and learning from the organization?
    - Ideas for Practice: Annual or semi-annual interviews about employee engagement, motivation and alignment with purpose and vision of programs.
- Does the structure and roles of the organization best meet the needs under present conditions?
    - Ideas for Practice: Surveys to evaluate effectiveness of leaders and opportunities for personal growth and development. (e.g. Indicate your level of agreement with the following statements: (1) I understand my roles and responsibilities, (2) I am supported and encouraged to take on new roles and responsibilities to support my professional growth).

However, new evaluation and assessment tools should be aligned with the organization's values and have pathways for informing future decision-making. This is an ongoing challenge for UAEO and other urban agriculture educators who might be struggling to differentiate between what the organization holds valuable in competition with what its community partners might hold valuable or what is required for reporting to funding agencies.

Considerations for STEM Reform and Urban Agriculture Education.

While it might be pragmatic to align urban agriculture educational programming with language of STEM reform in collaborations with public schools, it is important to consider that many food system activists may be working against a functionalist, corporatized vision of education. One that focuses on narrowly serving the interests of business elites, who are possibly set to benefit from a better trained work force to fill technical careers that produce products and services that might be implicated in the further degradation of natural resources and continued marginalization of humans. For example, UAEO was born out of visions of alternative models of learning and participation in modern, post-industrial democratic societies and working towards education for sustainable development and land resource management. Furthermore, UAEO founders initially aligned its educational and agricultural practices with science disciplines like agroecology (Wezel et al. 2009) that have grown out of resistance to technocratic, energy and capital intensive forms of agriculture which, despite increased productivity, have insufficiently supported progress towards social justice or sustainability. Over time, the organization remains distanced from these more political forms of activism and engagement, though they have dedicated efforts to address racial injustice in the food system more recently. Apolitical approaches to urban agriculture education may appease potential funders and offer new and robust development opportunities, but may also limit capacity for more holistic transformation. An apolitical stance may have played a critical role in longevity of the organization and its ability to present unifying messages around good food education, without specifying exactly what is meant by good food. However, Meek and Tarlau (2016) caution the enactment of apolitical garden-based education in their articulation of a framework for Critical Food Systems Education. We believe these tensions are embedded in urban agriculture education programs writ large and are relevant to all practitioners of urban agriculture and STEM education. In understanding the practice of balancing fertility in urban agriculture education settings, we suggest that at the least, over time many volunteers and employees may bring expectations of more transformative brands of food justice education. For other practitioners, these sorts of programmatic decisions and approach to engaging in urban agriculture education may position an organization within a matrix of being more or less transformational. Regardless of this position, we argue that the practice of building trusting and mutually beneficial relationships with stakeholders is key to program longevity and must be examined by organizational leaders and researchers.

7.12 Afterword

This chapter is now undergoing final revisions nearly 9 years after initial drafts and data collection were conducted. Upon revisiting this chapter, we saw artifacts of unexamined privileges and biases that warrants substantial re-orientation to approaches to urban agriculture and STEM education. This is especially necessary in circumstances wherein a predominantly white-led organization such as UAEO serves historically marginalized BIPOC learners. In re-imagining the stories that we describe herein, the easiest remedy would have been to include the voices and lived experiences of the BIPOC students and communities instead of centering the voices of the white urban agriculture educators that fit into an archetype of seemingly well-intentioned individuals "serving" a community while also materially and financially benefiting from participation in deficit narratives. A key, yet, underexamined factor in the community ecosystem described above is institutionalized racism that is embedded into the history and present realities of the food system (Penniman 2018). Pursuits in urban agriculture and STEM education, including this chapter, should acknowledge this as a starting point for future progress.

References

Blair, D. (2009). The child in the garden: An evaluative review of the benefits of school gardening. *Journal of Environmental Education, 40*(2), 15–38.

Cho, J., & Trent, A. (2006). Validity in qualitative research revisited. *Qualitative Research, 6*(3), 319–340.

Coffey, A., & Atkinson, P. (1996). *Making sense of qualitative data: Complimentary research strategies*. Thousand Oaks: SAGE Publications, Inc.

Coleman-Jensen, A., Rabbitt, M., Gregory, C., & Singh, A. (2020). *Household food security in the United States in 2019*, U.S. Department of Agriculture, Economic Research Service
Lave, J., & Wenger, E. (1991). *Situated learning: Legitimate peripheral participation*. New York, NY: Cambridge University Press.
Maanen, J. V. (1988). *Tales of the field: On writing ethnography*. Chicago: The University of Chicago Press.
Meek, D., & Tarlau, R. (2016). Critical food systems education (CFSE): Educating for food sovereignty. *Agroecology and Sustainable Food Systems, 40*(3), 237–260. https://doi.org/10.1080/21683565.2015.1130764.
Olson, S., & Labov, J. (2014). *STEM learning is everywhere: Summary of a convocation on building learning systems*. Washington, DC: National Research Council of the National Academies. DOI: https://doi.org/10.17226/18818
Penniman, L. (2018). *Farming while black: Soul fire Farm's practical guide to liberation on the land*. Chelsea Green Publishing.
Stake, R. E. (1995). *The art of case study research*. Urbana-Champaign: Sage Publications, Inc.
Tamboukou, M. (2008). Re-imagining the narratable subject. *Qualitative Research, 8*(3), 283–292.
Tan, E., Calabrese Barton, A., Kang, H., & O'Neill, T. (2013). Desiring a career in STEM-related fields: How middle school girls articulate and negotiate identities-in-practice in science. *Journal of Research in Science Teaching, 50*(10), 1143–1179.
Weaver-Hightower, M. B. (2011). Why education researchers should take school food seriously. *Educational Researcher, 40*(1), 15–21.
Wenger, E. (1998). *Communities of practice: Learning, meaning, and identity*. New York, NY: Cambridge University Press.
Wenger, E., McDermott, R., & Snyder, W. M. (2002). *Cultivating communities of practice: A guide to managing knowledge*. Boston: Harvard Business School Publishing.
Wezel, A., Bellon, S., Doré, T., Francis, C., Vallod, D., & David, C. (2009). Agroecology as a science, a movement and a practice. A review. *Agronomy and Sustainable Development, 29*, 503–515. https://doi.org/10.1051/agro/2009004.
Winne, M. (2010). *Food rebels, guerilla gardeners, and smart-cookin' mamas: Fighting back in an age of industrial agriculture*. Boston: Beacon Press Books.
Zeidler, D. (2016). STEM education: A deficit framework for the twenty first century? A sociocultural socioscientific response. *Cultural Studies of Science Education, 11*, 11–26.

Chapter 8
The USDA Future Scientists Urban Agriculture Program

Craig Wilson, Carolyn Schroeder, and Timothy Scott

Abstract The decisions that are made today about the future of agriculture are critical. Over the next 50 years there will need to be produced as much food as has previously been produced in recorded history. This will only happen if students enter science, technology, engineering, agriculture and mathematics (STEAM) fields. Only about 1/32 of the Earth's surface is suitable for farming. Water resources are being used at unsustainable rates. The students of today are going to have to solve and ameliorate these problems and many more as adults. However, the majority of these students live in urban environments with little to no idea of the source of their food. Students need to understand the complexity of the living world and to study the functions and processes of organisms and how they interact in and with the environment. The USDA Future Scientists Urban Agriculture Program, established in 2009, is a way to make that connection by having them study and research in a USDA People's Garden and then to develop their own garden at their school and, perhaps, at their homes. It is helping many of them to expand their sense of future science career options and to see Agricultural Science as moving beyond manual labor and the harvesting of crops from the field and into advanced research.

Keywords People's garden · Pollinator · Vegetable garden · Place-based outreach education · Experiential learning · Language arts · STEAM · Interdisciplinary

---

C. Wilson
Texas A&M University, USDA Southern Plains Agricultural Research Center (SPARC), College Station, TX, USA
e-mail: cwilson@science.tamu.edu

C. Schroeder
College of Science, Texas A&M University, College Station, TX, USA
e-mail: cschroeder@science.tamu.edu

T. Scott (✉)
Office of the Provost, Texas A&M University, College Station, TX, USA
e-mail: t-scott@tamu.edu

I. DeCoito et al. (eds.), *Teaching and Learning in Urban Agricultural Community Contexts*, Urban Agriculture,
https://doi.org/10.1007/978-3-030-72888-5_8

8.1 Introduction

When President Lincoln founded the United States Department of Agriculture (USDA) in 1862, he called it the *People's Department*. It was the first governmental department formed when 98% of the citizens (the People) were involved in some form of agriculture and 2% were not, roughly the reverse of today. In 2009, the Secretary of Agriculture Tom Vilsack began The People's Garden initiative. The program is an effort to challenge employees to create gardens at USDA facilities.

One such garden was created outside the USDA offices in College Station, Texas. Its design, implementation and educational outcomes on learners is the subject of this chapter. It combined a pond, a Texas pocket prairie, a pollinator garden and a Monarch Butterfly Waystation, information on which may be found at: http://monarchwatch.org/waystations/. There was also a vegetable garden, the latter created in collaboration with the USDA Farm Service Agency (FSA).

This USDA initiative coincided with the establishing of the Committee on Learning Science in Informal Environments (CLSIE) by the National Research Council (NRC), also in 2009. The brief of the CLSIE was to study the potential for science learning in non-school settings where people spend the majority of their time and to ascertain if science is actually learned through everyday experiences, in designed spaces, and in programs run by science enthusiasts. The committee found evidence that structured, non-school science programs can indeed feed or stimulate the science-specific interests of adults and children, may positively influence academic achievement for students, and may expand participants' sense of future science career options (Bell et al. 2009; Tai et al. 2006).

Importantly, health benefits accrue from all types of gardening. A meta-analysis of research examining the effects of gardening, including horticultural therapy, on health was conducted by Soga et al. (2017). Most of their studies came from the United States, followed by Europe, Asia, and the Middle East. Their findings suggest that gardening has several beneficial effects such as reductions in anxiety and depression. In short, gardening is beneficial for your health. One participant in a study in England commented that, "It made me feel brighter in myself." The project had introduced ornamental plants to thirty-eight previously bare front gardens within an economically deprived region of North England, UK. Four key themes emerged from the qualitative data analysis (interviews). Introducing plants elicited feelings related to motivation, relaxation, pride and positive emotions (Chalmin-Pui et al. 2021). In 2018 in the Republic of Korea, four researchers worked with maladjusted elementary school children to see if having them take part in the gardening activities of flower arranging, planting and pressing flowers would reduce their stress levels. It did so significantly (Jung Lee et al. 2018).

8.2 Design

A combination of wishing to be a part of the USDA People's Garden Initiative and to be able to use the garden for educational outreach resulted in the creation of the USDA People's Garden in College Station, Texas, by USDA employees and volunteers. It was considered a 'designed space' as defined by the U.S. National Research Council (NRC) 2009 Consensus Report as a space that can support science learning: "rich with real world phenomena, these are places where people can pursue and develop science interests, engage in science inquiry, and reflect on their experiences through sense-making conversations" (p. 3). A conscious decision was made to focus outreach efforts at the elementary level and on fifth grade students. This is a crucial age when research shows there is a gradual decline in children's motivation (Bathgate et al. 2014) for science. The garden was a motivational tool to open the students' eyes to the limitless possibilities available to them in the fields of science. Practical experiences in the garden allow for this to happen. It is important that children learn to see science not as a subject taught only in books in isolation in a classroom, but as a subject that is embedded in everything we do (Rahm 2002). Using the analogy of the trees that they will study in the garden, science has many branches and extensive roots.

The decision was taken to start small and, literally, to grow a garden that would have five components, the first being a pollinator garden. The area chosen was a utility easement beside the USDA offices that was overgrown and used as a dumping ground for old mattresses and bicycles (Fig. 8.1). It was an uninviting spot that needed to be reclaimed and repurposed.

The underbrush was removed, specimen trees left standing, sand trucked in to fill in what had been a drainage ditch and the area leveled (Fig. 8.2).

Fig. 8.1 Site selected for the USDA People's Garden

Fig. 8.2 Site after brush was cleared and sand added

Fig. 8.3 Trellis and first plantings installed

A small area was planted with butterfly attractant plants and shrubs like lantana (*Lantana horrida*) and verbena (*Verbena canadensis*) (Fig. 8.3). These are hardy and easy to establish. One goal was to use native plants and allow them to spread as groundcover to minimize the need for weeding. Pathways were created with coarse bark chips underlain with landscape fabric, also to cut down on maintenance. This must be augmented annually as it breaks down gradually.

Next on the agenda was the creation of a Monarch Waystation, a pollinator garden meeting specific requirements of the Monarch Butterfly (*Danaus plexippus*). This insect engages in one of the world's greatest migrations and is a natural wonder. The garden is on their flight path in the fall from Canada and the Mid-West to Mexico and their return migration each spring, so establishing gardens in Texas is critical to sustain the '*Miracle Monarch Migration*' (Fig. 8.4).

As the Monarchs funnel through Texas on their way north during their spring migration, it is imperative that the adults find milkweed plants (Family: *Asclepias*) on which to lay their eggs as this is the only food for their caterpillars It is this migration that is under threat due to a number of factors, but the prime one is the loss of their natural habitat that includes milkweed plants. There are over 100 species of milkweed that grow throughout mainland USA and 30 species are found in Texas. The two most common species found growing in Texas are Antelope Horns (*Asclepias asperula*) and Green Milkweed (*Asclepias viridis*), and the one most often planted in gardens is Butterfly Weed (*Asclepias tuberosa*). All three were planted in the Monarch Waystation and other milkweed species were added to help contribute to monarch conservation.

The garden sits alongside a shopping mall and there is a grass buffer zone between the two. The latter could be termed a 'green desert' (Fig. 8.5). This is an area that is very demanding of resources. It has an extensive irrigation system of water sprinklers that are activated whether water is required or not in both winter and summer. Often excess water is wasted and can be seen flowing away along the roadside. The grass is also heavily fertilized and takes a tremendous amount of labor when mowed weekly or bi-weekly. It may look neat, tidy and manicured but it is a

Fig. 8.4 Rain collection system and milkweed blooming

Fig. 8.5 The dormant grasses in pocket prairie next to green desert

barren habitat that is home to almost no wildlife. This location provides an excellent opportunity to educate both students and the general public about the choices that they make.

A conscious decision was made to turn part of the garden into a 'pocket prairie' to serve as a comparison for students. The latter is an attempt to recreate a small area in an urban environment to exemplify how the Texas prairie landscape looked before colonization and development. Advice was solicited from experts at a local company that produces native seed, including both grasses and wildflowers that require no extra watering once they have established their roots. No fertilizer is needed and the plants are naturally disease- and bug-resistant while the area provides ideal habitat for insects, birds and small mammals (Fig. 8.6).

Native grasses planted include eastern gamagrass (*Tripsacum dactyloides*), inland sea oats (*Chasmanthium latifolum*), little bluestem (*Schizachyrium scoparium*) and sideoats grama (*Bouteloua curtipendula*), the Texas state grass.

Native wildflowers were planted including American basket flower (*Centaurea americana*), gay feather (*Liatris mucronata*), lazy daisy (*Aphanostephus sp.*) and bluebonnet (*Lupinus texensis*), the Texas state flower.

Humans need to eat, too, so in collaboration with the Farm Service Agency (FSA) an area was set aside for a vegetable garden and tilled by FSA volunteers (Fig. 8.8). The goal was for the students to get their hands dirty and to see how edible plants are grown. Metaphorically, the children are cultured and grow in understanding as they cultivate vegetables. Research by Leuven et al. (2018) suggests that the exposure to vegetables generated by school gardening programs may enhance students' willingness to taste and to increase their daily intake of vegetables in the long term.

Fig. 8.6 Texas wildflowers blooming in the pocket prairie

Poulson et al. (2014) point out that urban agriculture began as early as the 1890s, had a resurgence during the two World Wars, and started to become popular again in the 1970s as a way to supplement the family food budget and out of concerns for the adverse effects of modern farming technologies on the environment and health. The vegetable garden is a valuable component of the People's Garden, as it models for students how a relatively small area, formerly 'a food desert', can be turned into a flourishing oasis that is rich in physical, psychological, relational and educational resources (Fig. 8.7).

The final component of the garden was the addition of a pond (Fig. 8.8). Since the land had formerly been a drainage ditch, and an undeveloped area of the garden held rain after a storm, it was decided to erect a simple clay dam, creating a pond. Kneeling boards were placed around the edges for ease of access and study by students. The deeper part was planted with native aquatic plants and, where it merged into wetland, the shallow area was made accessible by addition of a wooden pallet walkway. Arkansas rough rice (*Oryza sativa*) is planted annually around the edges.

The garden was specifically designed to emphasize agricultural science, a critical yet often-overlooked field of study. Perhaps its profile and importance might be

Fig. 8.8 Pond within the USDA People's Garden

elevated were the common acronym of STEM expanded to read: Science, Technology, Engineering, Agriculture and Mathematics or STEAM. Within the far-reaching field of agriculture, it is possible to cover many elements of the Next Generation Science Standards (NGSS) (NRC 2013), addressing Interdependent Relationships in Ecosystems, Cycles of Matter and Energy Transfer in Ecosystems, and Ecosystem Dynamics, Functioning and Resilience. Importantly, the students will learn by doing, an educational philosophy promoted by Dewey (1938) but now challenged by teaching to the test philosophy and practice. The USDA People's Garden provides a science-rich environment for educational outreach to benefit both students and the general public.

8.3 Implementation

A collaboration was formed with four local elementary schools, each with high enrollments of historically underserved students. A *Future Scientist Activity Notebook* was developed by the Program Director in collaboration with local elementary teachers. It was to be used by students to collect information through making detailed observations and taking accurate measurements in the garden and then recording these in their own notebook. The notebook had been tested and refined the previous year with other elementary students. The activities focus on urban agriculture, entomology, wildlife and botany. A grant was obtained from the USDA/

Fig. 8.7 Vegetable portion of USDA People's Garden

Hispanic Serving Institutions National Program (HSINP) that provided funds for two teachers from each of the four collaborating schools, eight in total, to be trained in the use of the Junior Master Gardener (JMG) Wildlife Gardener curriculum (Seagraves 2004) and to use activities from *Access Nature* (Almeras and Heath 2001). These eight teachers then visited the USDA People's Garden with their students to see an example of the type of garden they could create at their schools with grant funding. Each school then developed and built a USDA Future Scientists Garden. The teachers and students were helped in developing the garden by their school principal, teachers, local volunteer Master Gardeners, Master Naturalists and parents, with additional input from the JMG Program Coordinator, native wildflower seed experts and local USDA/Agricultural Research Service (ARS) research scientists.

In 2010–2011, four schools brought their fifth grade students on four separate field trips to the USDA People's Garden, just a few miles from each school. Students spent a whole day in the People's Garden. They were divided into eight groups of 6–8 students under the leadership of a USDA/ARS or FSA staff volunteer and rotated between eight separate activities:

1. HAIKU POEMS. Creativity is a huge part of a scientist's arsenal of skills, as is writing. Alberts (2010) suggests connecting science and literacy, using poetry as

one means to do so. Similarly, Walders (2000), a poet and former teacher, wove poetry into all areas of her curriculum, especially the sciences, and witnessed the power of the right poem at the right moment. Such creative writing skills are often overlooked but speak to the value of a good foundation of basic learning being laid at an early age. In order to scaffold a poetry activity, an example of a poem is provided and each student then chooses where to compose their poem, be it sitting in the garden, perhaps among the wildflowers in the pocket prairie, or any location that will allow them to observe nature at work, with these observations providing inspiration. For example, fifth grade student Austin Hamm penned the following haiku poem (Fig. 8.9):

Monarchs beautiful
Travel through the clear blue sky

Fig. 8.9 Haiku illustration

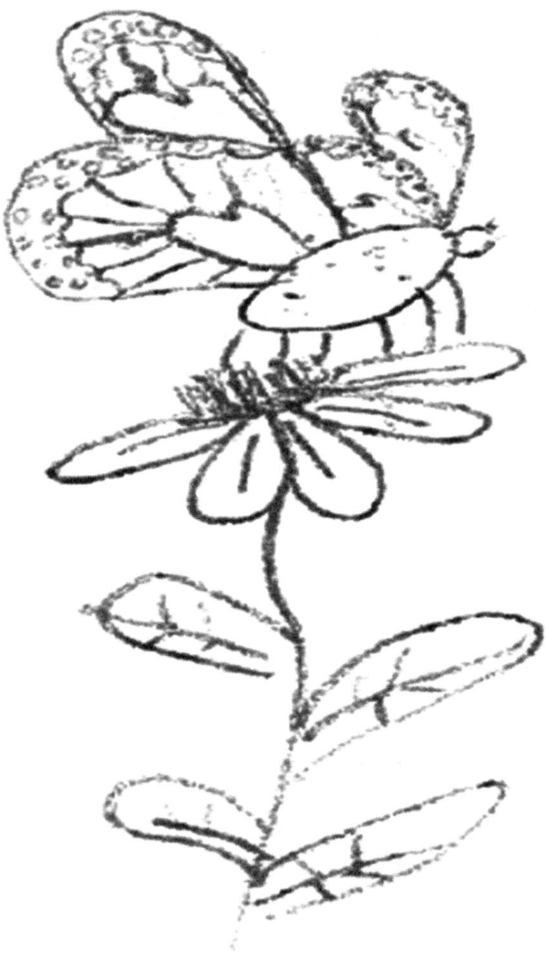

And, no-one knows why!

2. LEAF COLLECTION (Fig. 8.10). In an activity like this it is important to give the students a focus, thus they are asked to focus on the leaf edges. There can be nine classifications of leaf edges, including entire, undulate, serrate, doubly serrate, dentate, crenate, lobed-pinnately, parted-pinnately, or lobed-palmately, as described in the book, *Taxonomy of Flowering Plants* (Porter 1967). Students collect one of each type of leaf edge and these are then pressed within their notebook.

3. MEASUREMENTS. Too often, students are taught the units of measurement but have little understanding of the concept. In this activity they use non-standard units like paces to measure aspects of the garden, e.g., length and width, and then use meter sticks to better understand the standard metric units. Students are then given the freedom to make any measurements they choose within the garden, such as the diameter of a flower or the length of a grasshopper's antenna.
4. GARDEN INSECTS. A USDA/ARS research scientist or biological technician comes across from the local research lab and volunteers his/her time to show the students how to use 'bug-sweep' nets, how to transfer their catch into a clear plastic bag, and then how to identify the insects. The students are also able to carefully observe insect behavior in the gardens.
5. FLOWER OBSERVATION. The students are asked to observe, make drawings and identify all types of flowers. There are Texas native wildflowers in the pocket prairie and surrounding woodland, nectar source flowers in the pollinator garden, and even peanut flowers in the vegetable garden. They are provided with a laminated flower identification chart.

Fig. 8.10 Leaf collection activity

6. POND STUDIES. Students are drawn to the water. They may net a native fish, either mosquito fish (*Gambusia*) or sunfish (*Lepomis macrochirus*), to observe and take measurements before drawing (Fig. 8.11). They take the temperature of the pond water and compare it to the air temperature. They make notes on what is needed for the fish to survive in their aquatic environment and can identify the aquatic plants using guides such as *Pond Life* (Reid 2001) in the Golden Books series.

7. IDENTIFYING BUTTERFLIES and TAGGING a MONARCH. The gardens are a haven for butterflies as well as other insects. The students come to understand that the adage, "*Plant it and they (the insects) will come*", is true. They observe insects in the garden and note that different species have different host flowers. For example, the Gulf Fritillary (*Agraulis vanilla*) lays its eggs on the passion vine (*Passifloraceae incarnata*) while the monarchs lay their eggs solely on the milkweed leaves. During the monarch's fall migration, the students become a part of the citizen science project out of Kansas University and are able to catch a butterfly and attach a small identification number to the hind wing (Figure 8.12). More information about monarch identification and tagging can be found at the following website: https://monarchwatch.org/tagging/

8. VEGETABLE GARDENING. This activity is led by Farm Service Agency (FSA) personnel who till an area ahead of time and create rows so that each group of eight students has their own row in the plot. Students are allowed to handle and try to identify various seeds and then shown how to plant seeds and seedlings and are allowed to plant their own, such as peanuts (*Arachis hypogaea*), carrots (*Daucus carota*), onions (*Allium cepa*) and the 'eyes' of potatoes (*Solanum tuberosum*). The grand finale is when each student is allowed to pull up a pre-planted carrot, wash it and eat it to taste the freshness.

Fig. 8.11 Student drawing of pond life

Fig. 8.12 Monarch identification and tagging

8.4 Educational Outcomes for Learners

As suggested by Honey et al. (2014) the program integrates content and skills. Each group undertakes four activities in the morning, recording their observations and data. There is a lunch break and they complete four remaining activities in the afternoon (Fig. 8.13). The primary goal is to engage teachers and students outside the classroom, as the school gardens become outdoor laboratories. They benefit from interacting with experts from horticulture, botany, wildlife and entomology. They come to understand the value of outdoor learning. For example, students at one of the schools produced a video where they follow the pollinating activities of a bee. It is set to music and shows the students studying in the garden while a caption runs beneath the visual stating: "*The garden is a great place to go and see, in real life, all the things we learned about in class. One example is how flowers are pollinated by bees. We are lucky to have such a garden at our school.*" The students had ownership of the project and, as found by O'Neill (2005), were better engaged in the learning environment as a result. A garden allows teachers and students, as observed by Thorp (2005), to connect to each other, to connect to the healing rhythms of nature, to feed hungry children, and to introduce innovation and creativity into a mandated curriculum.

The four schools involved in the project are located in areas with large populations of historically underserved students. Seventy-seven percent of the students in this district identify as Black or Hispanic. The gardens are the only places that provide insight for these students to get to understand where and how their food is produced. We have come full circle since the late nineteenth century when gardens were used to integrate school and society, theory and practice (Kohlstedt 2008). The

Fig. 8.13 Field day at the USDA People's Garden in College Station, Texas

USDA Future Scientists Urban Agriculture Program uses the environment and the garden to help deepen students' understanding of basic science phenomena.

The USDA People's Garden is an extension of an existing program, the USDA Future Scientists Program, that targets science teachers, specifically their teaching skills, science content knowledge, and use of hands-on, inquiry-based science activities (Scott et al. 2011). The garden component ensures that all students have the opportunity to learn and benefit from significant involvement in creating, maintaining and conducting research in their own school garden where they have ownership. Learning by doing, termed 'practice' in the NGSS (NRC 2013) was found to be an essential, but underutilized, method of learning (Moye et al. 2014). Agriculture is a huge industry and the students' appreciation of gardening and wildlife could position them to consider Agricultural Science as their major, should they choose to go to college.

There is anecdotal evidence that the gardens are having a positive impact. Docherty (2014) cited the following quotes from a principal, teacher and fifth grade student:

> Edward Fellows, Fannin Elementary principal said, "Dr. Wilson has worked with the students through … creation of butterfly gardens and research in the USDA People's Garden. Each time [Wilson's] with kids, he is looking to engage students in their thoughts and willing to go the extra mile while offering experiences to kids in an unorthodox manner."

> Theresa Robinson, a fifth grade teacher at Johnson Elementary who brings her students to the garden said, "When they return to Johnson their senior year of high school, I have

noticed an increasing number of my former students actually going to college to major in some sort of science-oriented subject."

A fifth grader who interacted with Dr. Wilson, Cody Wall, completed a degree in Wildlife and Fisheries Management at Texas A&M University and attributes his success and extends his gratitude to Wilson. "I always remember his approach, 'The best way to learn is to get your hands dirty,' … as it's true that I have learned the most out in the field doing research rather than being nose-deep in a book."

8.5 Conclusion

The decisions that are made today about the future of agriculture are critical. In 2009 the chief executive of the Australian Commonwealth Scientific and Research Organization, Dr. Megan Clark suggested that over the next 50 years, it is estimated that we will need to produce as much food as has previously been grown in recorded history. Only about 1/32 of the Earth's surface is suitable for farming. Water resources are being used at unsustainable rates. As adults, today's students are going to have to solve and ameliorate these problems and many more. Work and research in the garden help students to understand the complexity of the living world as they study the functions and processes of organisms and how they interact in and with the environment. Zarger (2008) claims that gardening cultivates a 'sense of wonder' for the workings of the natural world (p. 8). There is even a garden in space within the Zvezda Service Module of the International Space Station (ISS). Psychologists at the Russian Institute for Biomedical Problems have concluded that tending the garden has beneficial calming effects on their cosmonauts (Johnson 2010). Anecdotally, it has been observed that both astronauts and cosmonauts are drawn to experiments where plants are being studied compared to inanimate research projects.

The book, *Gardening for Nutrition* (2014), can provide a helpful template for starting and developing a school garden. It is designed for use in Florida but has wider application, with supporting data to show that student achievement scores have improved as a result of gardening. Similar results are recorded in California, along with promoting healthy lifestyles, environmental stewardship and social development. But as found by Burt et al. (2019), successful implementation of a school garden program is predicated on engaging key stakeholders like administrators, teachers, parents, the community at large, and garden coordinators. The USDA Future Scientists Urban Agriculture Program managed to engage multiple stakeholders to create a successful garden with the wider goal to renew the thirst for knowledge in students who, when younger, were renowned for asking questions to the point of distraction: Who? Why? What? Where? and When? (Wilson et al. 2010).

Meyer and Lederman (2013) note that teachers have to plan for creativity. What better way than for a teacher to plan and build a garden with students. Gardening is one focus of the NGSS (NRC 2013). A school garden provides a way to explore the integration of science and engineering practices, disciplinary core ideas and crosscutting concepts. Other core ideas can also be covered and studied in the garden:

ecological interdependence, growth and development of organisms, structure and function, adaptation, and the environmental impact of human activity. The NGSS strategy of science notebooking is well suited for use in the garden, integrating science and literacy. Students learn the language of science by making observations and recording data (Fulton 2017). Another NGSS strategy that lends itself to use in the garden is science talks where the students actively listen, ask questions, and build on the ideas of others.

Urban gardening programs engage students in real-world, practical learning that can motivate them to pursue studies in the sciences. Not everyone is academic and learning in the outdoors has the opportunity to re-engage with students who have not been successful in traditional settings. It may appeal to diverse students from different racial and ethnic backgrounds and preliminary studies from a research project, Science in the Learning Gardens (SciLG) by Williams et al. (2018) provide support for the notion that learning in school gardens has the potential to promote science equity. Students experience different ways of learning science that may lead to improved success in science and feelings that they can become part of the wider scientific community now and in the future.

Supportive environments such as a garden need to be developed to nurture student creativity and to release their imaginations for, as Albert Einstein said, "Imagination is more important than knowledge. Knowledge is limited. Imagination encircles the world" (Viereck 1929). Integrated STEAM learning, outdoors in an urban garden, is part of the answer to re-engaging students in science. It is a way to get them involved experientially, to encourage them to ask questions once more, and to interest them in finding imaginative solutions to the world's agricultural problems that lie ahead. Their answers and solutions will be critical.

References

Alberts, B. (2010). Prioritizing science education. *Science, 329*, 748–749.

Almeras, B. G., & Heath, D. J. (2001). *Access nature*. Washington, DC: National Wildlife Federation.

Bathgate, M. E., Schunn, C. D., & Correnti, R. (2014). Children's motivation towards science across contexts, manner of interaction, and topic. *Science Education, 98*, 189–215.

Bell, P., Lewenstein, B., Shouse, A. W., & Feder, M. A. (Eds.). (2009). *Learning science in informal environments: People, places, and pursuits*. Washington, DC: National Academy Press.

Burt, K., Lindel, B., Wang, J., Burgermaster, M., & Feras, J. (2019). A nationwide snapshot of the predictors of and barriers to school garden success. *Journal of Nutrition Education and Behavior, 51*, 1139–1149.

Chalmin-Pui, L. S., Roe, J., Griffiths, A., Smyth, N., Heaton, T., Clayden, A., & Cameron, R. (2021). "It made me feel brighter in myself."- the health and Well-being impacts of a residential front garden horticultural intervention. *Landscape and Urban Planning, 205*.

Dewey, J. (1938). *Experience and education*. Toronto: Collier-MacMillan Canada, Ltd..

Docherty, D. (2014, September 16). *Professor leads future scientists. The Battalion*. Retrieved from http://www.thebatt.com

Florida Agriculture in the Classroom, Inc. (2014). *Gardening for nutrition*. Gainesville: FAITC.

Fulton, L. (2017). Science notebooks as learning tools lessons from a multi-year professional study group offer insights on getting the most out of science notebooks. *Science and Children; Washington, 54*(6), 80–85.
Honey, M., Pearson, G., & Schweinngruber, H. (2014). *STEM integration in K-12 education: Status, prospects, and an agenda for research*. Washington, DC: National Academy Press.
Johnson, N. B. (2010). Religion, spirit and the idea of garden. *Religious Studies Review, 36*, 1–4.
Jung Lee, M., Oh, W., Soon Jang, J., & Young Lee, J. (2018). A pilot study: Horticulture-related activities significantly reduce stress levels and salivary cortisol concentration of maladjusted elementary school children. *Complementary Therapies in Medicine, 37*, 172–177.
Kohlstedt, S. G. (2008). "A better crop of boys and girls": The school gardening movement, 1890–1920. *History of Education Quarterly, 48*(1), 58–93.
Leuven, J., Rutenfrans, A., Dolfing, A., & Leuven, R. (2018). School gardening increases knowledge of primary school children on edible plants and preference for vegetables. *Food Science and Nutrition, 6*(7), 1960–1967.
Meyer, A. A., & Lederman, N. G. (2013). Inventing creativity: An exploration of the pedagogy of ingenuity in science classrooms. *School Science and Mathematics, 113*, 400–409.
Moye, J. J., Dugger, W. E., & Starkweather, K. N. (2014). "Learning by doing" research – introduction. *Technology and Engineering Teacher, 74*(1), 24–27.
National Research Council (NRC). (2013). *Next generation science standards*. Washington, DC: The National Academies Press.
O'Neill, T. (2005). Uncovering student ownership in science learning: The making of a student created, mini documentary. *School Science and Mathematics, 105*, 292–301.
Porter, C. L. (1967). *Taxonomy of flowering plants*. Caldwell: The Blackburn Press.
Poulson, M. N., Hulland, K. R. S., Gulas, C. A., Pham, H., Dalglish, S. L., Wilkinson, R. K., & Winch, P. J. (2014). Growing an urban oasis: A qualitative study of the perceived benefits of community gardening in Baltimore, Maryland. *Open Anthropology, 36*, 69–82.
Rahm, J. (2002). Emergent learning opportunities in an inner-city youth gardening program. *Journal of Research in Science Teaching, 39*, 164–184.
Reid, G. K. (2001). *A guide to pond life*. New York, NY: St. Martin's Press.
Scott, T., Wilson, C., Upchurch, D., Goldberg, M., & Bentz, A. (2011). The USDA and K-12 partnership: A model program for federal agencies. *Journal of Natural Resources and Life Science Education, 40*, 29–35.
Seagraves, R. (Ed.). (2004). *Junior master gardener: Wildlife gardener*. College Station: JMG Kids.
Soga, M., Gaston, K. J., & Yamaura, Y. (2017). Gardening is beneficial for health: A meta-analysis. *Preventive Medicine Reports, 5*, 92–99.
Tai, R. H., Liu, C. Q., Maltese, A. V., & Fan, X. (2006). Planning early for careers in science. *Science, 312*, 1143–1144.
Thorp, L. (2005). A season for seeds. *Culture and Agriculture, 27*, 122–130.
Viereck, G.S. (1929, October 26). *What life means to Einstein*. Saturday Evening Post. Retrieved from http://www.saturdayeveningpost.com
Walders, D. (2000). *Poetry and science education*. Office of Educational Research and Improvement, Washington DC. (ERIC Document Reproduction Service No. ED 463 946)
Williams, D. R., Brule, H., Kelley, S. S., & Skinner, E. A. (2018). Science in the Learning Gardens (SciLG): A study of students' motivation, achievement, and science identity in low-income middle schools. *IJ STEM Ed, 5*, 8. https://doi.org/10.1186/s40594-018-0104-9.
Wilson, C., López, J. D., & Scott, T. P. (2010). "Who ate our corn?" We want to know and so should you! In R. E. Yager (Ed.), *Science for resolving issues/problems*. Arlington: NSTA Press.
Zarger, R. (2008). School garden pedagogies: Understanding childhood landscapes. *Anthropology News, 49*, 8–9.

Chapter 9
Forging the Farm-To-School Connection: Articulating the Vision Behind Food-Based Environmental Education at The Dalton School

Kevin Slick and Mila Tewell

Abstract The study and pursuit of "Urban Agriculture" takes the form of food-based education at the Dalton School (a private K-12 institution in New York City founded in 1919 by Helen Parkhurst). Working with and within the constraints of an urban environment, the Dalton School utilizes a flexible and adaptive approach to food-based environmental education in order to provide the basis for more holistic approaches to environmental sustainability. Dalton actively engages students to think of themselves as stewards of the earth and their community, and both curricular and extra-curricular activities aim to do so. From the creation and maintenance of a "Community Supported Agriculture" (CSA) enterprise, to a "Pop Food" course, to a senior history elective on the study of food production, to a grade-wide emphasis on sustainability in the second grade, and the recent construction of an in-house teaching kitchen, Dalton provides multiple opportunities to engage ecological and environmental issues through the study of food. The ultimate goal is a curriculum that is vertically aligned and introduces developmentally appropriate curricula in terms of both content and practice and thus creates continuity between the grade levels. Additionally, a food-based approach to environmental education can demystify traditional notions of nature (and the preconceptions that attend to it) and open up a pathway for examining issues of personal identity and ethnicity, social justice and racial inequality and providing an anti-racist framework of analysis and action.

Keywords Sustainability · Food · Food justice · Environmental education · Teaching kitchen · Social and racial inequality · Nature · Ethics

K. Slick (✉) · M. Tewell
The Dalton School, New York, NY, USA
e-mail: ks2@dalton.org; mtewell@dalton.org

I. DeCoito et al. (eds.), *Teaching and Learning in Urban Agricultural Community Contexts*, Urban Agriculture,
https://doi.org/10.1007/978-3-030-72888-5_9

9.1 Introduction: Creating a Food-Based Curriculum

At the Dalton School in New York City (a private K-12 institution founded in 1919 by Helen Parkhurst), our curriculum is intentionally action based, asking students to engage their academic subjects both in the classroom and in the field. As the school reached its centennial, many faculty members actively sought to re-examine and reaffirm the pedagogical principles of our guiding philosophy, articulated in our founding document, *Education on the Dalton Plan.* In this way, we are re-engaging the practices of student-centered and project-based learning (Parkhurst 1922) and actively supporting areas of student interest as they pertain to the educational experience and beyond. Within this larger context, "food" has emerged as a locus of active learning and direct engagement. In our food-based curricula, the interdisciplinary opportunities abound. Throughout our K-12 program, courses and student clubs are emerging that focus on food. In our First Program (K-3), students address issues of sustainability in general, visit farms, learn about the growing cycle, and create their own farmers' market in the spring. In the middle school, a "Science Technology Engineering Arts and Math" (STEAM) summer camp provides students with the opportunity to take a class on design and production of green walls and grow their own food. The high school welcomes students with a variety of examples including a Food Club, which oversees the work of our student-run Community-Supported Agriculture (CSA) serving 100 Dalton families, and addresses issues of food security in the New York metropolitan area, through political action and social service. From biology to history to sustainable engineering, food is increasingly a topic of intellectual discourse and student-initiated action at the Dalton School.

At the Dalton School, urban agriculture finds its clearest expression in food-based environmental education, and the overarching goals can be summed up in a few significant ways. For one, this approach to education promotes the development of personal responsibility and a heightened awareness of the sustainable stewardship of the land, food production, and eco-literacy. Furthermore, food-based education provides an entry into the complex and rich ecological history of humans' engagement with the natural world. Following the words of Aldo Leopold from his *A Sand County* Almanac (1949), at the heart of this experience is an appreciation of a land ethic as a living ethic: "an ethic dealing with human's relation to land and to the animals and plants which grow upon it." In this way, food-based environmental education has become more and more incorporated into our students' curricular and extracurricular life at Dalton, and will, we hope, serve as the basis for the cultivation of essential life values and the promotion of sustainable well-being. The purpose of this chapter is to highlight what has been happening at Dalton in the realm of food-based environmental education and to indicate some plans for future growth.

It is difficult to identify the exact moment in which food-based environmental education began to gain traction at the Dalton School. The learning environment at Dalton is best characterized as a dynamic one in which students and faculty actively share ideas and interests. Topical concerns play a great role in the everyday discourse, and the invitation to innovate and to express freely ideas and interests inheres in the "Dalton Plan" as codified by Dalton's founder Helen Parkhurst in her *Education on the Dalton Plan* (1922). The intense interest in food emerged roughly five years ago and reflected the broader cultural concerns with food production, consumption, and preparation. As food became more of a concern within popular culture and the everyday experiences of our students, there emerged a growing curiosity to learn more about the world of food and to better understand the character and specific nature of our food pathways. A colloquium on food offered numerous opportunities to engage in food-themed issues ranging from keynote talks given by Mario Batali, Dan Barber, Eric Schlosser, and representatives from the Slow Food movement to workshops led by farmers, food purveyors, butchers, activists, and food writers. In the wake of the success of the colloquium, Kevin Slick developed a senior elective on food, entitled *You Are What You Eat: A Critical Investigation Into Food*, to reflect student interest and to further the understanding of food as the site of intense ecological, agricultural, cultural, political, and economic forces. The words of Wendell Berry (noted farmer, writer, and activist) provided us (and continue to do so) with a crucial and endearing point of departure: "A significant part of the pleasure of eating is in one's accurate consciousness of the lives and the world from which food comes" (Berry 2009, p. 234). Even the most cursory examination of popular media from books to blog, from television shows to the proliferation of farmers' markets and farm-to-table restaurants would suggest we live in an age in which we have arrived at a kind of culinary enlightenment. The pleasures of eating can be profound. However, we also live in an age in which growing numbers of the population in the United States (and elsewhere in the world) can be defined as both overweight *and* undernourished (Claasen et al. 2016). The abundance of seemingly diverse consumer goods to be found in the average supermarket belies the centrality of corn and soy as main ingredients. Variety gives way to uniformity. "Organic" operates as a potent signifier of pastoral values and earth-friendly practices, but the reality speaks to a far less idyllic paradigm of industrial production. In this way, food production is intimately linked to environmental degradation (Ritchie 2020).

The overarching goals of the aforementioned senior elective on food are quite simple: to achieve a greater awareness of the source and character of our food in order to better understand the environmental impact of our choices. We are constantly being told what to eat and how much, but this information is often conflicting, misleading, and just plain strange. Inspired by the many writings of Wendell Berry, Michael Pollan's *Omnivore's Dilemma* (2007) offers one of the seminal efforts to unearth the origins of our daily sustenance; indeed, to develop a more accurate consciousness of food culture, food production, and food pathways. Using his own personal experience as the basis to support his assertion that "the way we eat represents our most profound engagement with the natural world," Pollan (2007, p. 10) proposes to follow the food chains that sustain us. From the industrial, to the pastoral, and to contemporary efforts toward hunting and gathering, Pollan's work explores each food pathway in equal measure and posits the radical distinctions between them. However, Pollan was quick to remind us of Berry's elemental truth behind every practice: "eating is an agricultural act" (Pollan 2007, p. 11). To this end, we are encouraged to consider even the Twinkie: even it finds its humble origins on the farm (in the form of corn, many times removed from the actual plant).

9.2 Putting Food at the Center of Environmental Education

A critical examination of food begins with the most elemental questions. What is food? How have we come to eat what we eat? What are the various factors influencing our choices? What informs our decision making process? If we return to Pollan's *Omnivore's Dilemma* (2007), we realize these questions have no ready and singular answer, and the difficulty of answering these questions speaks to a unique condition of our historical moment. To be certain, we are in need of guidance. In short, food-based environmental education uses food to explore the historical, ethical, scientific/ biological, cultural, aesthetic, political and economic forces shaping what we eat (Gordon and Hunt 2019). This approach to education and the experience of being on the farm afford a context in which we can understand food production, preparation, and consumption as elements of a rich and complex ecosystem. To be alive is to engage in the food system. How we engage and why we do so is a matter of consciousness and conscientiousness. When it comes to the choices we make when we eat, food-based environmental education helps us explore the convergence of the local and the global, urban and rural, and the past and the present. We can more readily highlight and explore eating as a profound engagement with (or alienation from) the natural world indicated by our food consciousness and current practices of production and consumption.

Our historical investigation into food is one of recent vintage: we live in an age of amazing food choices and consciousness. However, we examine how there is no singular shared attitude toward food: despite the attention paid to the local, organic, and artisanal, popular attitudes toward food are heterogeneous and conflicting and even polarizing. Food, something so seemingly elemental, carries mixed connotations. Melanie Rehak (2015), in a recent article summed up this conundrum very effectively:

> To me, all of this signifies that we're living in a kind of post-food world … we're now living in an era when food, whether it's mass-produced or carefully grown on a tiny plot, always represents a larger concept – not just something to eat. What used to be a pile of vegetables is now an emblem, and we can never go back to the time before it was so.

Furthermore, conscientious eating is paramount but must be done critically. The *locavore* ethos has had great success and has helped bring about a shift in consciousness, but it is also important to rethink the farm-to-table impulse to include a more expansive view of the ecological setting of food production. Dan Barber's *The Third Plate* (2014) prompts a more nuanced investigation of the component parts of a truly sustainable food system which in turn is best understood as a web of relationships within an integrated whole encompassing growers and consumers, natural systems and cultural practices. In this way, we emphasize the "culture" in agriculture as a unified system of thought and memories, limits and actions.

In the same way that our emphasis on global citizenship prepares our students for an ever-evolving and exciting future, an emphasis on the "local" can also encourage and promote a deep understanding of the individual's role in a larger community and society. Food-based learning centered on the farm or other immediate means of food production, the epitome of the local, can help support the development of a strong sense of civic identity and responsibility, as well. The concept of sustainability encourages students to think holistically and critically about the environment and their role as conscientious consumers and producers. As a result, students can then see themselves as agents of change and effective decision-makers. In short, everyone eats and the food system is complex, and the emphasis on local production is an added feature of sustainability education. The *Farm-Based Education Association* expresses many of these beliefs below:

> We believe that farm-based education is among the most effective and promising forms of environmental, experiential, and place-based education because of the innate ability in all people to connect to farms.
>
> We believe that farm-based experiences promote life values by relating to the social, moral, cognitive, and emotional aspects of the human experience.
>
> We believe that hands-on learning on farms builds confidence, self-awareness, and individual and collective responsibility which leads to the sustainable stewardship of our world.

In *The Pleasures of Eating*, Wendell Berry (2009) provides the reader with seven principles (p. 232) (serving as guidelines) for navigating contemporary food choices which will be outlined in the remainder of the chapter. Berry inveighs against the industrial food system in this essay (and many others), but his intentions are to liberate us from what appears to be a fixed dependence on impersonal and (seemingly) monolithic systems of food production. There is great optimism in his work, and he seeks to demonstrate alternatives. Therefore, these guiding principles ground us again inescapably in the world, and our responses in turn indicate how we seek to understand and use the world by the choices we make.

1. **"Participate in food production to the extent that you can … "**

The emphasis on food and food production is becoming increasingly popular and seen as vital to a student's awareness of self and social and historical place. One recent study of farm-to-school programs indicated the success of these programs in terms of supporting an understanding of long-term sustainability of food systems, raising awareness of a variety of health and environmental issues, and promoting experiential learning (Joshi et al. 2008). Additionally, in a study designed to assess its program's effectiveness, the Edible Schoolyard maintained the centrality of the kitchen, garden, and farm as the site for cultivating a better sense of how the natural world sustains us and promotes the overall environmental and social well-being of the school community. To quote Alice Waters, restaurateur and founder of the Edible Schoolyard, "our program places food at the center and has students helping to feed each other. These children learn mutual respect from sharing meals, they learn self-respect from learning how to prepare meals and they learn respect for the planet from learning how to grow food in an ecologically sound way" (Center for Ecoliteracy 2003, p. 2).

The vast majority of food we consume is produced on farms located well outside of an urban setting, and not all farms are the same. The range of agricultural practices is staggering, and our main level of participation in the food system is as consumers. With students serving as farmers, an emphasis on farm-based education allows for the greatest possible awareness into the essential features of production. At Dalton we acknowledge the inherent limits on our capacity to replicate the dominant food systems, but we still engage directly in horticultural practices as educational methods. Indeed, the constraints of urban living can provide opportunity in the form of rooftop gardens, community gardens, and home gardens, and these features serve as crucial points of entry toward the fulfillment of the larger goal of understanding food production. Dalton offers a terrace garden of about 500 square feet and students routinely and increasingly utilize this site as a learning opportunity. For one, students in the middle school have developed a "gardening club" during which they do the work of tending to the garden. These activities range from selecting and ordering seeds, weeding, planting, harvesting, and maintenance. The garden provides a great opportunity for thinking about seasonality and what is appropriate for our space and climate zone. Students in "Gardening Club" keep records and measure the progress of plant growth in relation to rainfall, temperature, and season. Moreover, the garden is evolving into an effective demonstration garden as students studying colonial America in the fourth grade have planted examples of Native American agriculture in the form of the "Three Sisters" and students in high school environmental science have dedicated beds for exploring food production by way of Dan Barber's *The Third Plate* (2014). As students make use of these resources more actively and effectively, the vitality of the garden will continue to grow as an educational medium.

The design and construction of "Growing or Green Walls" has also been explored in an independent study by juniors in our Environmental History class. Dalton will now offer a student led class on green wall design and construction for middle school students in our STEAM summer camp. Faculty from the science department – in sustainable engineering and biology – are joining the history faculty in supporting our student teachers, with the goal of filling Dalton with student built green walls in the coming years.

2. **"Prepare your own food. This means reviving in your mind and life the art of kitchen and household … "**

An awareness of food production coupled with an understanding of food preparation in its most tangible and vivid forms will broaden an understanding of natural processes. Cooking (facilitated by the attendant culinary skills) is an act of transformation. As Michael Pollan (2013) puts it so eloquently in his latest work, *Cooked,*

> [h]andling these plants and animals, taking back the production and preparation of even just some part of our food, has the salutary effect of making visible again many of the lines of connection that the supermarket and the 'home-meal replacement' have succeeded in obscuring, yet of course never actually eliminated. To do so is to take back a measure of

> responsibility, too, to become, at the very least, a little less glib in one's pronouncements...Especially one's pronouncements about the 'environment,' which suddenly begins to seem a little less 'out there' and a lot closer to home. For what is the environmental crisis if not a crisis of the way we live? (p. 21)

The Dalton offers a "Pop Food" class in an effort to explore and elucidate the food consciousness suggested by authors like Pollan, Waters, and Barber. To this end, "The High School Pop Food" curriculum is divided into a series of modules that are offered at different times throughout the year – Saturdays, afternoons, evening, and early mornings. The goal is to allow students several options to integrate the modules into their existing schedule. Each module culminates in a pop up restaurant. "Bread Module" supports a one-day bakery; "After the Harvest" results in jars of jam, pickles, and invitation-only lunches; "Coffee Module" opens a coffee shop in the school lobby every morning for a week; and the "Restaurant" accommodates about 100 people in for a sit-down dinner.

By gearing the curriculum toward a culminating event, we hope to provide something more than a knife-skills class, and more of a holistic food experience – students see where the food comes from, how to cook it, and how to offer it to those who are keen to eat. In this way, cooking becomes something not abstract, but a hands-on real-time experience, just as the performing arts curricula are connected to a performance or team sports to an actual game. It also means the students become responsible for a wide range of demands such as publicity, menu design, transformation of the cafeteria space into a restaurant, ordering rentals, securing bags or other necessary packaging, and creating an on-line ordering website.

The goal of food-based environmental education is not necessarily to create future farmers or chefs (although that may happen). Rather, this approach to learning allows for the inculcation of an enlightened food consciousness and the development of essential culinary, economic, and social skills. The corollary between a critical inquiry into food and an understanding of the scientific principles found in the study of biology, chemistry, physics, and a food-based program will assist in the promotion of this interdisciplinary approach to learning. For instance, in ninth-grade biology, the Dalton School uses a workshop on the making of sauerkraut as an opportunity to further an investigation into the principles of fermentation. These workshops both enhance and help develop an understanding of the properties of lacto-fermentation and the microbial world and provide a vivid and exciting hands-on opportunity. This is in keeping with a greater emphasis on the use of food as the basis of student-led inquiry into the physical sciences, and increased efforts are being made to incorporate food into lab practicals.

3. **"Learn the origins of the food you buy and buy the food that is produced closest to your home. The idea that every locality should be, as much as possible, the source of its own food makes several kinds of sense. The locally produced food supply is the most secure, the freshest, and the easiest for local consumers to know about and to influence."**

This principle situates food within a broader social dimension, and thus allows us to see the local cultural spaces, landscapes, histories, collective experiences, and traditions associated with food. In many ways, this speaks to the heart of the emerging food movement characterizing our present.

However, there is a central irony at work here, as well. A focus on local production and consumption gives us insight into the growing inequities associated with the contemporary food system and its pathways. Even with greater attention placed on local production, we become keenly aware of the asymmetries of distribution and access. For instance, a 2008 study by the Department of City Planning in New York City indicated that "nearly three million residents live in 'high need' neighborhoods" (Gottlieb and Joshi 2010, p. 41). Furthermore, this report "utilized a supermarket index that correlated neighborhoods with the highest levels of diet-related diseases and the most limited opportunities to purchase fresh foods" (Gottlieb and Joshi 2010, p. 41). Issues of food justice, according to Gottlieb and Joshi (2010), can and should be a significant feature of a farm-based educational experience:

> Issues of race, ethnicity, class, and gender are at the forefront of a social justice agenda. When viewed through the food justice lens, they are also a key part of how we talk about food. Historically, food groups have struggled to effectively address the contentious issues associated with questions of class, race, ethnicity, and gender….The language of food justice ensures that these core issues are not ignored but rather placed at the center of the discussion regarding how, by whom, and to what ends the food system is transformed. (p. 229)

In an effort to address, mitigate and/or eliminate the effects of asymmetrical food sovereignty and access, students at Dalton are taking action in the form of "Generation Citizen," a student led class, teaching "action civics." Students choose a local issue to address in a year-long activist campaign, reaching out to community leaders and service providers to effect needed change. The most recent campaign was to address food (in)security in New York City. Students reached out to officials with SNAP and NGOs advocating for the food insecure, to shelters to offer classes on cooking and nutritious foods to homeless children, to New York City schools to explore expanding knowledge of and access to good food, and to the private sector, both markets and produce suppliers to find creative ways to address this long term issue. The class' work will be continued by a newly created "Food Club," the stated mission of which is "to educate our community on issues of food security and sustainability and to promote the availability of affordable, nutritious food to all New Yorkers."

Keeping within the realm of food justice, Dalton's "Fair Trade Store" is an outgrowth of our school-wide Global Initiative, the History of Food elective, and another history elective on The Developing World. The latter's curricular focus on development and the role of social entrepreneurship led students to explore the role of fair trade in global trade, and potential partnerships with producers around the world. Our first two products are chocolate produced by a Grenadian farmer cooperative – marketed to students at local events, and a "supergrain" indigenous to Senegal and currently being developed on a local level by a women's cooperative in the south of that nation. Working closely with partner farmers and producers, students address real world business problems and use a range of skills to develop open markets to these entrepreneurs.

4. **"Whenever possible, deal directly with a local farmer, gardener, or orchardist."**

In order to achieve the best possible impression of the season, this is an essential practice. Understanding how to do this effectively is another outcome of food-based education: the synthesis of knowledge and flavor and a direct connection between production and consumption. This point becomes particularly poignant and enduring when the students are the farmers or deal directly with farmers.

Dalton's Community Supported Agriculture (DCSA) is a high school student initiative that contracts with local farmers (from the greater New York area) to deliver farm fresh produce to 100 Dalton families throughout the school year. Created two years ago, DCSA has become a thriving, fully student run operation, serving as a community-building program and educational opportunity. The success of the program at Dalton inspired the DCSA leaders to write a "how to" manual and engage other schools in New York City, who are now partnering with Dalton students to develop their own CSAs. Our contacts with farmers have inspired further classroom and field-work opportunities.

5. **"Learn, in self-defense, as much as you can of the economy and technology of industrial food production. What is added to food that is not food, what do you pay for these additions."** and
6. **"Learn what is involved in the *best* farming and gardening."**

Principles 5 and 6 speak to the apparent "efficiency" and "productive capacity" of the industrial food system which in turn gives the impression of progress: we are surrounded by abundance and readily accessible and relatively inexpensive food. But what are the real costs of the efficiency and abundance provided by the industrial food system? In his essay *Agricultural Solutions for Agricultural Problems* (1979), Wendell Berry is addressing the hidden costs of the industrial food system: "costs" that are further obfuscated (perhaps deliberately) by our detachment and disassociation from the means of production. Soil erosion, soil compaction, soil and water pollution, ecological degradation and deterioration, and the use of increasing amounts of chemical fertilizer to offset the loss of soil fertility and to combat the growing threat of pests and diseases are but a few of the negative externalities of the dominant food system.

At the heart of this consideration is an opportunity for moral clarity. The costs of the industrial food system are well documented. Our system of abundance is sustained by exploitation of workers and animals and resources (Committee on a Framework for Assessing the Health, Environmental, and Social Effects of the Food System et al. 2015), and farm-based environmental education strives to highlight the true inefficiencies of the current dominant system and introduces us to viable alternative models.

During the spring semester, Dalton's environmental science class devotes considerable time and effort to an understanding of best practices in agriculture, defined by sustainability, diversity, and balance, leading to an understanding of the greater ecosystem of food. Students in the class read Jonathan Safran Foer's *Eating Animals*

(2009). An avowed vegetarian, Foer offers both a philosophical statement for vegetarianism and a critique of the modern industrial food chain. Foer's work allows students to address the philosophical dimensions of the general desire to eat animals and the resulting systems developed to further and sustain this craving for animal flesh. Foer devotes considerable attention to the document of the horrors of the "factory farm" (emblematic of the industrial food system), but he never loses sight of the centrality of death in the carnivore's diet. According to Foer (2009),

> Rationally, factory farming is so obviously wrong, in so many ways. In all of my readings and conversations, I've yet to find a credible defense of it. But food is not rational. Food is culture, habit, and identity. For some, that irrationality leads to a kind of resignation. Food choices are likened to fashion choices or lifestyle preferences – they do not respond to judgments about how we should live. And I would agree that the messiness of food, the almost infinite meanings it proliferates, does make the question of eating – and eating animals especially – surprisingly fraught... Food is never simply a calculation about which diet uses the least water or causes the least suffering. And it is in this, perhaps, that our greatest hope for actually motivating ourselves to change lies. In part, the factory farm requires us to suppress conscience in favor of craving. But at another level, the ability to reject the factory farm can be exactly what we most desire. (p. 263)

Although Foer's work can be read as a manifesto on vegetarianism, the rigor of his argument encourages a conscientious consideration of the lives of the animals to be consumed. While his work *is* an argument for vegetarianism, it is "also an argument for another, wiser animal agriculture and more honorable omnivory" (Foer 2009, p. 244). To achieve this honorable omnivory, to develop the accurate consciousness highlighted at the beginning of this presentation, requires a direct engagement with the realities of life and death on the farm. To do so will make us greater moral beings. The opportunity for an ensuing moral clarity works to mitigate the negative effects of the industrial food system. What is unique to food-based environmental education, perhaps, is the opportunity for an ever-present awareness of death. Our sustenance is predicated on production and destruction, and death is central to this equation. The treatment of animals and plants (and the people who tend to them) is not equal across the food system, and the spectrum runs from humane treatment to gross exploitation. Even a strictly vegan diet requires death through cultivation of land, and eradication of animal habitat. Students actively engage with these issues and encounter ethical issues in the context of a science class.

7. **"Learn as much as you can, by direct observation and experience if possible, of the life histories of the food species."**

This general theme has significant representation within the curriculum. As mentioned earlier, the environmental science class at Dalton includes a large unit on the study of food in its production and link to ecological effects. Dalton also offers a variation on the core eleventh-grade world history curriculum (a study of the twentieth century) in the form of a course devoted to the study of the events of the twentieth- and twenty-first centuries through an ecological lens. The foundational logic of this course includes a definition of human ecology in which we explore the complex relationship between human beings and their natural, social, and built environments. In this way, a solid evaluation and understanding of ecology and environmentalism relies on a keen perception of human behavior and the myriad relationships between the natural world and broader economic, political, social, and cultural forces. An environmental and ecological assessment of world history is really a broad frame/ lens through which we examine human activity in a larger ecological setting: action unfolds in various contexts. All human activity has an ecological impact, and all human behavior has an ecological basis, and this idea will serve as our guiding logic as we explore the events and developments of the twentieth century into the twenty-first. To this end, the course explores the role of land use, the struggle for energy, water rights and access, food shortages, deprivation, and the subsequent consequences of collectivization and rapid industrialization. Many of the environmental issues of the day have a direct antecedent in the events of the early to mid twentieth century (and often earlier), and attention to ecological developments provides the overarching narrative for our investigation.

The study of world history provides a wonderful opportunity to address environmental issues, but there is real concern on local activism and engagement throughout the Dalton curriculum. A focus on time and place provides a pathway into significant ecological developments and a history of land use – "of the ways in which human minds and hands have worked in tandem with natural opportunities and constraints" (Ryden 2011, p. 37). By engaging the landscape (through food, experience, history, and decoding traditions as markers of ecological possibility) and its history, we can discern and witness the transformation of the 'natural' world.

9.3 Food-Based Focus on the Natural and Applied Sciences and STEAM Subject Areas

The following list offers curricular guides for Dalton educators engaged in the incorporation of a food-based focus into STEAM subject areas:

- Exploration of water use and conservation (with an opportunity for water quality testing)
- Demonstration of alternative energy production from the sun and the wind
- Demonstration of innovations in engineering
- Demonstration of sustainable design and performance
- Demonstration of organic growing and composting (and the role of microorganisms)
- Data gathering on energy use and production
- Using the site and its physical plant as educational tools

A vertically aligned curriculum provides structure and occasion for exploring issues in a sustained manner across the K-12 experience. The goal is to introduce developmentally appropriate curricula in terms of both content and practice, and to create continuity between the grade levels. Some of the most significant work being done at Dalton can be found in the youngest grades, and this sets the stage for the development of a successful environmental and ecological awareness. For instance, a year-long theme for the second grade is a focus on sustainability. Students in this grade address issues pertaining to sustainability through a variety of activities (including trips to rooftop gardens and urban farms). Each student participates in a composting workshop and learns about "nature's recyclers" as a corollary to their own activities. Recycling and reducing waste are a major focus of the curriculum, and the second graders have taken the lead in managing and calculating the recycling efforts of their building (which house grades K-3). This helps reinforce their keen sense of place and personal responsibility, and an understanding of the sustainability features of Little and Big Dalton, the role of waste and resource management, the idea of composting, and a knowledge of best practices contributes to this awareness, as well. One activity to help further an understanding of the importance of recycling takes the form of food waste management. Students weigh their leftovers after lunch for a week or so in an effort to become conscientious eaters and reduce waste.

Activism around environmental issues starts early at the Dalton School, and every week second graders offer gentle messages of sustainability and responsibility via the public address system in an effort to promote awareness and encourage student action. The following are examples of their messages:

Use the blue recycling bin to recycle all kinds of plastic like yogurt containers and drinking cups. Make sure you empty the containers and rinse them out first.

When you or your parents go to the grocery store, remember to bring reusable bags with you. If you forget, try to reuse the paper or plastic bags from the store.

Remember that recycling is not just something we do at Dalton. Make sure you follow all the recycling rules at home and help out your family by teaching them the green tips you've learned.

This is just one of the many ways students in the First Program engage and understand environmental issues in a hands-on manner. A series of recent polls indicated that Americans on average do not routinely cite environmental concerns as one of the most pressing problems facing the country, but the energy and commitment to environmental change exhibited at the First Program is definitely not your average response to the ecological challenges facing all of us. This important work needs to be engaging and fun. This was evident in the Farmers Market run by two kindergarten classes last year as the culmination of a larger project centered on nutrition and a knowledge of food production and its impact on the environment. For one class, their market followed on the heels of a visit to the Union Square Farmers Market. For the other class, this was a further extension of two visits (one of which they were accompanied by a group of high school seniors) to Hilltop Hanover Farm in Yorktown, New York where they witnessed first-hand the dynamics of a working farm. The farm-to-Dalton connection could not be made clearer or more exciting.

9.4 Creating Spaces for Food-Based Environmental Education

In the fall of 2019, the Dalton School completed an expansion and renovation of its main building (108 East 89th, NYC) to include new floors dedicated to STEAM activities. The new space provided a home for a robotics studio, collaborative work environments, a maker space, greenhouse and hydroponic spaces, a green roof, and a state-of-the-art teaching kitchen. The teaching kitchen in particular, as it pertains to food-based environmental education, has played a key role in advancing many of the ideas mentioned above. It has also proven to be a hub for all manner of dynamic educational activities, both intentional and unexpected, and has emerged as a space for an evolving set of pedagogical practices. Curricular offerings help to ground the teaching kitchen as an educational space, and these courses offer educational experiences and more. Two course offerings include "Culinary Fundamentals: Lunch" and "Food Systems: From Seed to Bread." In the first course, students study a different cuisine or cooking challenge oriented around cultural traditions, culinary techniques, nutrition, and environment each week. Then on Fridays, the class serves lunch to a segment of the Dalton community, "sharing both the food they've cooked and the social or culinary context that prompted that week's concept." Overall, the emphasis is on technique, nutrition, and an embrace of the way cooking employs artistic, cultural, historical and aesthetic choices to create food that reflects the particular vision of the chef. Students in "Food Systems: From Seed to Bread" become physically engaged in the precision craft of baking bread while developing a keen understanding of the relationship between cause and effect of what they are learning and bringing to practice. The course description offers the following, and it resonates clearly with the broader set of concerns and practices highlighted in food-based environmental education at Dalton and indicates a move toward a more comprehensive vision of food as the center of pedagogical concept:

> Once you instigate an exploration of the science, flavor, traditions, and almost endless variations, it's easy to make deeper connections to almost every other subject our students study. Students will live in an age of complex vertical food systems, and need to be prepared to make thoughtful, creative, informed decisions around issues of factory farming, sustainability, biodiversity, energy consumption, and food security. Bread today can be made from factory farmed grain into processed and preserved product, or through an artisanal process that involves wild wheat, active fermentation, hand crafting, and carefully baked techniques. We will look at old and new ways of making bread, and see what is to be learned from both. To that end, students will become part of a daily bread production. This will allow us to ask complex questions about the structure of this food system, to look for ways to take action towards improving our connection to it, and to ask broader questions about other culinary ecosystems.

What is immediately noticeable in the spaces in and around the teaching kitchen is the buzz of activity as students engage in a variety of tasks while seeming to erase the boundaries between the curricular and the extracurricular, the social and the academic. Although robotics, engineering, and computer science are part of the general curriculum at Dalton and adhere to our standard schedule, much of the work in these courses takes place after school hours, beyond the constraints of the traditional schedule. Food prepared by students in food-based courses is often available for students engaged in STEAM work, and the setting becomes one marked by conviviality and productivity. What seems evident here, in addition to a balance between fun and concentration, is the display of deeper learning:

> In many of the high schools we visited, much of the most powerful learning seemed to occur not in core classes, but rather at the school's 'periphery' – in electives, clubs, and extracurriculars. Hidden in plain sight, these peripheral spaces often had a very different 'grammar' than the one that usually dominated core classes. In these spaces, students had real choices, learning by doing was the norm, there was time to explore matters in depth, and students were welcomed as producers rather than receivers of knowledge. (Mehta and Fine 2019, p. 5)

With the teaching kitchen as a resource, the objective becomes leveraging the facility with the curricular opportunity in an on-going and expanded manner. The goal is to reposition a traditionally "marginal" activity like cooking at the center of the academic experience. The teaching kitchen in this sense became a laboratory for learning and exploration and cooking then becomes a formal investigative practice. The alliance between the existing senior elective, "You Are What You Eat: A Critical Investigation Into Food Production, Consumption, and Justice" and the kitchen space (and those faculty who use it strategically for educational purposes), enabled students to readily and actively explore aspects of contemporary food pathways and practices. To begin the culinary journey, students engaged in all aspects of bread-making to learn about process and sourcing. In this hands-on class, students milled heirloom grains from Anson Mills, used a ten-year-old starter/mother, kneaded the dough, and discussed the realities of the contemporary food chain. With bread as an example, the students explored the historical, ecological, ethical, scientific/biological, cultural, aesthetic, political and economic forces shaping what and how we eat. They followed the food chain from grain to crusty bread and thought about the artisanal and local as opposed to the industrial and commercial.

Another activity in the teaching kitchen had students exploring industrial meat production and consumption. The investigation entailed a consideration of the economic and cultural factors shaping modern meat consumption and the environmental impact of such practices. Students examined the ecological impact of beef production, consider its ecological footprint, and compared it to plant-based alternatives. To further illuminate the ideas under examination, a Dalton chemistry teacher made a guest appearance in the kitchen, as well. The chemistry teacher demonstrated the science/chemistry behind Chicken McNuggets (as a way to illustrate highly processed foods) while another faculty member made chicken meatballs inspired by the well-known chef Yotam Ottolenghi. Both were hands-on activities for students.

9.5 The Pivot Toward Food-Based Environmental Education as Anti-Racist Education

Even though Dalton suspended in-person experiences in the spring of 2020, the reflection on food and its place within the broader political, ecological, social, and racial context continued to shape curricular opportunities and engagement with social justice and environmental issues. The year 2020 also brought forth an urgent need to reassess the power dynamics and legacy inherent in predominantly white institutions, and for Dalton this reappraisal takes place on the level of curriculum and interpersonal interactions. For one, the Dalton History Department is dedicated to developing students who are informed global citizens, who can identify and challenge racial bias. We ensure that our students see aspects of their own identities in the curriculum and are empowered to interrogate both historical and contemporary narratives. To be certain, a food-based curriculum provides a point of entry into these considerations and provides the basis for understanding the complex web of factors and intersectional identities at work in any scenario. We urge our students to ask big questions and wonder about the dynamics of power that inform history, and the exploitation and active marginalization of individuals and groups around the world, particularly as it pertains to the growing inequities associated with the contemporary food system and its pathways. To this end, we want students to create and sustain a community of thinkers engaged in critical investigations of food production, consumption, and justice.

As Jonathan Safran Foer says in his recent article, "The End of Meat Is Here" (Foer 2020), "[i]f you care about the working poor, about racial justice, and about climate change, you have to stop eating animals." The incorporation of this work (and others like it) allows for an analysis of food production and consumption in the context of COVID, climate change, and the struggle for racial justice and brings home the very personal relationship we all have to these broader forces by way of the food choices we make. In this way, personal food stories and food pathways become the site of intersectional identities both historical and contemporary. What stories do the foods we eat tell us about ourselves? Why do we eat what we eat? Why are some dishes (or ingredients) important markers of our personal histories and identities? Michael Twitty's *The Cooking Gene: A Journey Through African American Culinary History in the Old South* and Genie Milgrom's *Recipes of my 15 Grandmothers: Unique Recipes and Stories from the times of the Crypto- Jews during the Spanish Inquisition* serve as an inspiration and framework for exploring the relationship between food and culture and identity. For Twitty, utilizing key ingredients such as rice, molasses, cornmeal, sweet potatoes, and beans and peas informed a food story that helped him recognize the culinary traditions of African enslaved people and their descendants who were often invisible in the narratives of Southern foodways and culinary culture (Twitty 2018). Gilmore writes of the Crypto-Jewish roots of her Roman Catholic family and culinary experiences growing up in Havana and Miami and the meals representative of this intersectional family identity (Milgrom 2020). Food stories in this sense are simultaneously global and local, complex and earthy, historical and contemporary. Students can use their own stories to position themselves within the curriculum as they consider family histories and traditions. And clear instances and insights into equity of access and patterns of food distribution and consumption become readily apparent.

The work of Chris Emdin proves especially valuable, germane, and urgent as we navigate the complex realities of intersectional identities and engage in anti-racist work (Emdin 2017, 2010). Emdin calls for empowering students to "step into their own voices" and calls for educators to signal a desire and commitment to cultivate an environment in which students can "show up as themselves" in class proceedings (Emdin 2017). Moreover, Emdin demands educators adopt pedagogical strategies to reach students who are not as engaged in classroom proceedings, often black, Indigenous and people of color (BIPOC). One pertinent strategy for doing so is by disrupting conventional narratives and deliberately engaging our students in discussion and debate around the construction of our disciplines. In this way, student agency and questioning become paramount as we all commit to equipping our students with the tools necessary to effectively interrogate their various fields of discipline and inquiry. Anti-racist work begins with calling into question and interrogating the assumptions and power dynamics that inhere in any academic discipline and its practices and practitioners. A genuine focus on student experiences becomes a pathway forward toward a more equitable and collaborative set of educational norms.

In the case of anti-racist work in environmental education, food is both ubiquitous and deeply personal and provides a meaningful and critical foundation for analysis of personal identity and broader systems. The work of William Cronon proves useful in bridging a variety of perspectives presented in this chapter and helps to synthesize and unify several strands. In *The Trouble with Wilderness; or, Getting Back to the Wrong Nature* (Cronon 1996), Cronon declares it is time to rethink the idea of wilderness. Cronon's overall objective is for us to (re)examine and then commit to a view of environmentalism that does not locate nature and the wild elsewhere. Rather, Cronon wants us to idealize and investigate critically the environments in which we live in an effort to produce an environmental ethic focused on the serious environmental problems we experience at home as opposed to just "over there." For Cronon, our views on "nature" and the "wild" are social constructions conditioned by centuries of changing views on landscapes and wilderness and the accompanying activities and associations. "Wilderness" as we know it is a profoundly human creation: "the creation of very particular human cultures at very particular moments in human history."

Cronon highlights the manner in which ideas and ideals about nature emerged in the nineteenth century that held the wilderness as the solution to modern culture. Romantic views of nature helped shaped the idea of the "wilderness experience" (especially with the emphasis on the "sublime"), and these ideas in turn influenced and were reinforced by ideas of the "Frontier." Turner's "Frontier Thesis" was central to the popularization of this view. The sublime and the frontier converged to remake wilderness and invest it with moral values and cultural symbols. Cronon examines a variety of Romantic writers (both American and European) for evidence of the articulation and promotion of these views on nature and the wild as sites of mystery and opportunity for rugged individualism with the attendant emphasis on white masculinity.

Throughout his piece, Cronon describes the religious antecedents of cultural notions of "nature" and the "wild" (drawing particular strength from an investigation of cultural readings of the Garden of Eden and Adam and Eve's expulsion from it into the unknown wilderness associated with Satan), and situates this religious view at the center of a more modern quasi-religious idea of "nature" as a place of purity and restoration. Much of modern environmental thinking, according to Cronon, fosters and maintains an erroneous view of "nature" as an ideal and a place to return to while nature is in itself a highly constructed and ever-present aspect of our daily lives. A truly effective environmental ethic would encourage us to see "nature" everywhere particularly in the foods we eat, the resources we use, our modes of transportation, and the consequences of our actions both here and over there.

It is that idea of focusing on environmental issues being found here and there, at home and further afield, that unifies the ideas behind a food-based environmental education. Again, Wendell Berry asserts that a "significant part of the pleasure of eating is in one's accurate consciousness of the lives and the world from which food comes." Situating ourselves at the center of a complex web of historical, ecological, ethical, scientific/ biological, cultural, aesthetic, political and economic forces opens up a space for seeing the environment, various systems and their impact, and personal choice and responsibility in a new light. We all eat, but eating may represent many things to many people. By examining the convergence of the local and the global, urban and rural, and the past and the present, we begin to see differently the choices we make when we eat. Eating is a profound engagement with (or alienation from) the natural world indicated by our food consciousness and current practices of production and consumption. A food-based environmental education empowers us to follow the food chain and see where it leads us. Our engagement with food, food culture, food politics, the history of food (production, distribution, and development), food preparation, and the actual cultivation of food is then historical, theoretical, textual, and practical. It also opens a space for action and activism and anti-racist work. A critical food-based educational approach has the cultivation of an informed and participatory citizenry at its core as we engage in issues of environmental sustainability, environmental racism, food justice, social (in)equality, and beyond.

9.6 Conclusion: Food-Based Education as the Basis for a New Environmental Ethic

Food-based environmental education proves flexible and adaptive and provides the basis for more holistic approaches to environmental sustainability. Dalton actively engages students to think of themselves as stewards of the earth and their community. These efforts are nascent in many ways, but Dalton's commitment to sustainability is genuine and evolving rapidly. Dalton is making concerted efforts toward helping students understand the impact and reason for one's actions and helping to bring about change (both local and global) through personal empowerment. In this way, sustainability becomes the basis of a lived experience at the Dalton School and one that will guide students (and faculty) toward a clearer consideration of their actions and their individual and collective responsibilities. To educate effectively for sustainability, we need framework and opportunity, and clear links to the curriculum.

When food-based environmental education succeeds, the expression of community through an understanding of personal responsibility is achieved. Aldo Leopold's (1949) idea of a "land ethic" captures this sentiment perfectly as he writes, "[t]he land ethic simply enlarges the boundaries of the community to include soils, waters, plants, and animals, or collectively: the land." Furthermore, "a land ethic, then, reflects the existence of an ecological conscience, and this in turn reflects a conviction of individual responsibility for the health of land." Leopold presents this synthesis of ideas at the conclusion of *A Sand County Almanac*, a collection of stunningly poetic descriptions of the natural world. It is this awareness of the interaction between "nature" and "culture" and our role as stewards of the earth that sustains farm-based education. The future appears uncertain, but this approach to education is one based on active intellectual and physical engagement, collaboration, personal and collective responsibility, and hope.

References

Barber, D. (2014). *The third plate: Field notes on the future of food*. New York: The Penguin Press.

Berry, W. (2009). *Bringing it to the table: On farming and food*. Berkeley: Counterpoint.

Claassen, M. A., Klein, O., & Corneille, O. (2016). Poverty and obesity: How poverty and hunger influence food choices. *European Health Psychologist, 18*(Supp), 337–339.

Committee on a Framework for Assessing the Health, Environmental, and Social Effects of the Food System; Food and Nutrition Board; Board on Agriculture and Natural Resources; Institute of Medicine; National Research Council; Nesheim, M. C., Oria, M., & Yih, P. T. (Eds.) (2015). *A framework for assessing effects of the food system*. Washington, DC: National Academies Press. https://www.ncbi.nlm.nih.gov/books/NBK305181/. https://doi.org/10.17226/18846.

Cronon, W. (1996). Wilderness. In *Uncommon ground: Rethinking the human place in nature*. New York: W.W. Norton.

Education for Sustainability: Findings from the Evaluation Study of The Edible Schoolyard. (2003). A report commissioned by the Center for Ecoliteracy. http://docslide.us/documents/the-edible-schoolyard-improving-behaviour-and-academic-results.html

Emdin, C. (2010). *Urban science education for the hip-hop generation*. New York: Sense Publishers.

Emdin, C. (2017). *For white folks who teach in the hood... and the rest of y'all too: Reality pedagogy and urban education*. New York: Beacon Press.

Farm-Based Education Association Website. http://www.farmbasededucation.org/page/our-guiding-language

Foer, J. S. (2009). *Eating animals*. New York: Back Bay Books.

Foer, J. S. (2020). The end of meat is here. *The New York Times.*

Gordon, C., & Hunt, K. (2019). Reform, justice, and sovereignty: A food systems agenda for environmental communication. *Environmental Communication, 13*(1), 9–22. https://doi.org/10.1080/17524032.2018.1435559.

Gottlieb, R., & Joshi, A. (2010). *Food justice.* Cambridge, MA: The MIT Press.

Joshi, A., Azuma, A., & Feenstra, G. (2008). Do farm-to-school programs make a difference? Findings and future research needs. *Journal of Hunger & Environmental Nutrition, 3*(2–3).

Leopold, A. (1949). *The land ethic. A sand county Almanac.* http://www.waterculture.org/uploads/Leopold_TheLandEthic.pdf

Mehta, J., & Fine, S. (2019). *In search of deeper learning: The quest to remake the American high school.* Cambridge, MA: Harvard University Press.

Milgrom, G. (2020). *Recipes of my 15 grandmothers: Unique recipes and stories from the times of the Crypto-Jews during the Spanish Inquisition.* New York: Gefen Publishing House.

Parkhurst, H. (1922). *Education on the Dalton plan.* New York: E. P. Dutton & Company.

Pollan, M. (2007). *The Omnivore's dilemma: A natural history of four meals.* New York: The Penguin Press.

Pollan, M. (2013). *Cooked: A natural history of transformation.* New York: The Penguin Press.

Rehak, M. (2015). What was the Hip Butcher? Looking at twenty years of American eating, from Big Macs to DIY Bacon. *Bookforum.* http://www.bookforum.com/inprint/021_04/13922

Ritchie, H. (2020). *Environmental impacts of food production.* Published online at OurWorldInData.org. Retrieved from: https://ourworldindata.org/environmental-impacts-of-food [Online Resource].

Ryden, K. (2011). The Handselled Globe: Natural systems, cultural process, and the formation of the new England landscape. In B. Harrison & R. W. Judd (Eds.), *A landscape history of new England.* Cambridge, MA: The MIT Press.

Twitty, M. (2018). *The cooking gene: A journey through African American culinary history in the old South.* New York: Amistad.

Chapter 10
Urban Beekeeping as a Tool for STEAM Education

Thomas Schmitt, Kristian Demary, and Noah Wilson-Rich

Abstract Urban beekeeping is an innovative tool with which to test hypotheses and practice concepts relating to STEAM (science, technology, engineering, art, and math) education. STEAM education uses art and design as an approach to STEM disciplines in a creative and engaging way. Here, we provide an overview of what urban beekeeping is, and how students of all ages successfully use bees as a means of project-based learning. We provide an overview of the uses of bees, their nest within the greater hive structure, their pathogens and parasites and their phenotypic gambit, with a special focus on practical applications. We delve into case studies of multiple, actual classrooms to demonstrate practical uses for bees with STEAM learning in elementary through advanced education. Bees are a viable means with which to advance STEAM education for learners of all ages, backgrounds, origins, nationalities, colors, races, and interests.

Keywords Bees · Beekeeping · Urban · Agriculture · STEM · STEAM · Geometry · Curriculum · Architecture · Children

---

T. Schmitt
Departments of Biology and Marine and Environmental Studies, Northeastern University, Boston, MA, USA

The Best Bees Company Inc., Boston, MA, USA

Urban Beekeeping Laboratory and Bee Sanctuary Inc., Boston, MA, USA

K. Demary
The Best Bees Company Inc., Boston, MA, USA

Urban Beekeeping Laboratory and Bee Sanctuary Inc., Boston, MA, USA

Liberal Arts, Massachusetts College of Art and Design, Boston, MA, USA
e-mail: kristian.demary@massart.edu

N. Wilson-Rich (✉)
The Best Bees Company Inc., Boston, MA, USA

Urban Beekeeping Laboratory and Bee Sanctuary Inc., Boston, MA, USA
e-mail: noah@bestbees.com

I. DeCoito et al. (eds.), *Teaching and Learning in Urban Agricultural Community Contexts*, Urban Agriculture,
https://doi.org/10.1007/978-3-030-72888-5_10

10.1 Introduction

To find sustainable solutions to real-world problems, students must be able to think and model across different disciplines and in terms of complex systems. Science, technology, engineering, art, and mathematics (STEAM) education uses art and design as an approach to STEM (without art) disciplines in a creative and engaging way (http://stemtosteam.org/). By taking a STEAM approach, teachers can engage divergent thinkers and spur innovations necessary for a strong economy (House Resolution 319 2011; Maeda 2013). Nature can inform us about applied geometry and be a source of inspiration for innovative designs and new technologies.

Field-based observations of the natural world, the environment, and human activities are powerful motivators for encouraging scientific learning. Urban ecology is the study of the interactions of species and their urban environment. This takes into account how humans are altering the landscape. Urban areas can be viewed as a heterogenous mix of land uses. Thus, urban areas can be viewed as living laboratories to study how land-use change affects biodiversity. This multidisciplinary systems-approach addresses Next Generation Science Standards (NGSS) key concepts (LS2.A, LS4.D and ESS3.C) of how living and nonliving components of the ecosystem interact to determine biodiversity. Urbanization has been associated with biodiversity loss for many biota (Czech et al. 2000; Czech 2004). In the United States, 275 species are listed as endangered due to urbanization (Czech 2004). However, invertebrate abundance and diversity can be high in urban green spaces, from green rooftops to domestic urban gardens (Jaganmohan et al. 2013; reviewed in Jones and Leather 2012).

Honey bees are a good springboard for STEAM learning while addressing and practicing the NGSS across the different disciplines: Life Sciences, Physical Sciences, Earth Sciences, and Engineering, Technology and Applications of Science. Managed honey bees can be viewed either in an observational hive in the classroom or outside in a traditional Langstroth hive. Where this is not an option, wild bees can be observed beyond the walls of traditional classrooms. Viewing bees both in their hives and in nature can foster curiosity and promote the process of science by students designing and conducting experiments.

Urban beekeeping, along with urban agriculture, can help people living in cities feel more connected with nature. This connection allows for opportunities to learn about the interconnectedness of plants, pollinators, and people. Project-based learning with bees can encourage critical thinking skills, problem solving, decision making, inspire art and technology, and ultimately result in the pursuit of STEAM training, education, and careers.

For younger learners, bees can be used to teach: animals are living, have life cycles, interact with the environment, and have specific habitat needs. For older learners, bees can be used to teach about: classification by their physical characteristics, their stages of development including metamorphosis, how changes in their environment can lead to either their death or relocation, and how bees cause changes in their environment (via pollination). Honey bees are a good tool to address the

NGSS: Life Sciences (LS2: Ecosystems and LS4: Biodiversity and Humans) because they are uniquely adapted to collect pollen and nectar and in doing so, they are important pollinators to over 70 crops. Honey bees (*Apis mellifera*) contribute an estimated $14.6 billion per year to the American economy (Morse and Calderone 2000). As a model system for research, their contributions to our understanding of the natural world are invaluable. Yet, honey bees are dying. The stability of the world's food production remains in question. Of the world's most important monoculture crops, 68% require animal pollination, and honey bees contribute a major portion of this (Klein et al. 2007).

One way to foster curiosity and critical thinking skills is to view honey bees in their hives. An observation hive is a unique tool for researchers and students to engage in collecting data from live honey bees in their natural, crevice habitat. These enclosed indoor structures are made of wooden, rectangular frames with glass on two sides. Nearly everything that occurs inside the honey bee hive is visible through the glass, including intimate interactions between and among the bees. Hundreds to tens of thousands of live honey bees thrive year-round inside this wood and glass structure, living on natural beeswax foundation incorporating hexagonal architecture. These bees access the outside via a tube to collect nectar and pollen. The information gained from the use of observation hives in scientific research is invaluable. In a classroom or museum setting, observation hives can allow students to learn about tessellations, the life-history of honey bees, the behavior of bees (such as the waggle dance, a Nobel-prize winning discovery for Karl von Frisch in 1973), the chemistry of honey, photoreception, and presence of disease and parasites (such as *Varroa* mites). With project-based learning, the students are able to make observations, form hypotheses, design their own experiments, analyze their results, and present their data visually. Students can graph the productivity (pollen and honey), fitness and fecundity (brood, adults, queen cells), and health (disease/parasite abundance) of the observation hive over time.

In this chapter, we discuss four cases in which bees can be used as a platform for STEAM learning. Case 1 is the use of managed honey bees in an observation hive in the classroom. Case 2 is the use of managed honey bees in a traditional Langstroth hive. Case 3 is the study of wild bees in Citizen Science projects. Case 4 is the study of wild bees at either pollinator gardens, insect or bee hotels (Fig. 10.1). We will use examples from three Massachusetts Public Schools: Mission Hill Elementary, Fenway High School, and Hingham High School. In addition, we will use examples from higher education with students from Northeastern University and Massachusetts College of Art and Design. Here, we provide specific examples of how teachers and students have put the urban beekeeping case study to practice both inside and outside of the classroom. Rather than posit how one could hypothetically test hypotheses relating to STEAM education using bees, we detail how actual students have already begun this exploration (Fig. 10.1). These examples are only the initial steps toward establishing urban beekeeping as a model system for STEAM practice. As more citizen scientists engage with urban bees, more observations will come to light, leading to an inevitable marriage between STEAM learning and practice, and bees, for future generations to come.

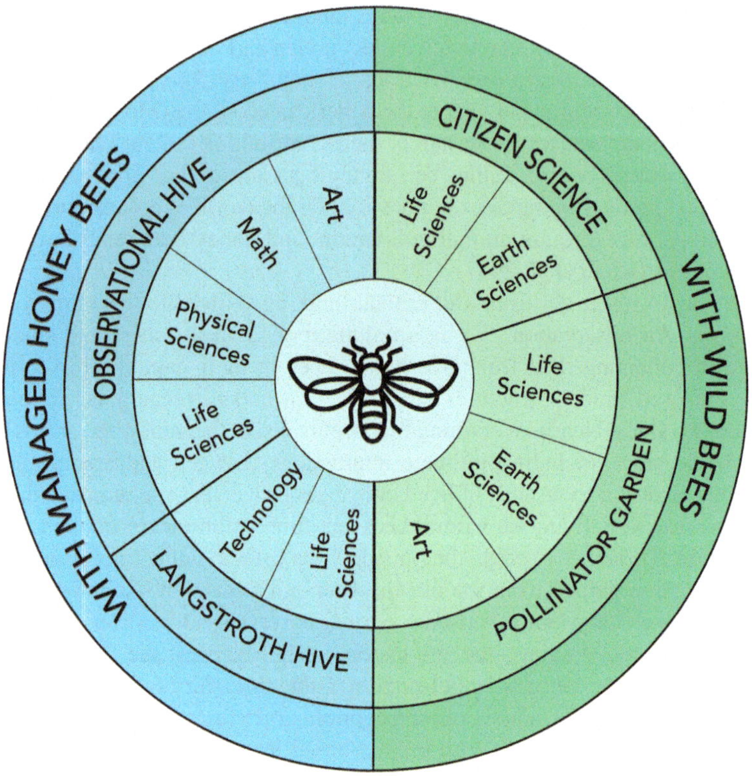

Fig. 10.1 Using honey bees to teach science, technology, engineering, art and math (STEAM). (Figure created by Jessica Lindsay)

10.1.1 CASE 1: Observation Hive in the Classroom

Honey bees in an observation hive do require maintenance. Classroom Hives is a nonprofit resource helping to guide teachers through the process of building, maintaining, and cleaning observation hives (http://www.classroomhives.org/). In addition, Classroom Hives has available teacher resources (http://www.classroomhives.org/?page_id=182).

10.1.1.1 Case Study: Mission Hill Elementary School

The Boston Public School system has multiple schools with classroom observation beehives. In 2001, the Mission Hill Elementary School installed an observation beehive used by teacher Ms. Jenerra Williams' third and fourth grade classes. Ms. Williams first has the students read the 4H Youth Project Book 1, *The Buzz About Bees: Honey Bee Biology and Behavior* (Fisher et al., 2014). Ms. Williams bases her

"Nature Science Curriculum" on the scientific method, starting with teacher-led observations of bees inside and outside of the beehive. Students engage with bees by observation through a glass, observation hive. Students then work together in groups to have discussions and ask questions about their bee observations. They come up with suppositions as a group – basic hypotheses to explain the mechanisms driving their bee observations. From here, Ms. Williams guides the young students' ideas to design experiments for testing their hypotheses.

One example of project-based learning is an experiment investigating how the age of worker bees determines the tasks they perform. Honey bees change their behavior with age, a process called temporal polyethism. Students first make and record observations, noticing that not all workers perform the same tasks. They begin by asking questions, such as, "Does the age of the bee determine what tasks she performs?" See Table 10.1 for an overview of the tasks of worker bees at different ages. Students can observe the behavior and tasks of different bees and display this visually with a chore chart (Fig. 10.2). In Ms. Williams' classroom, students design experiments to test the hypothesis that bee age determines behavior. Beyond her class, any classroom can work with a beekeeper's help to mark newly emerged bees with non-toxic paint on their thorax. This allows students to follow that age cohort of bees over time. This method is also safe for students to do, if comfortable, because honey bees are soft and unable to sting during their first 24 h after emergence as an adult. Results from this experiment facilitates the understanding that these students have of bees, while utilizing the scientific method as a learning framework.

Ms. Williams highlights the actions of bees (via pollination) to our food security with her Pollination Lesson Plan (outlined below).

Table 10.1 Honey bee temporal polyethism: the median age (in days) that workers perform certain tasks (Data from Winston 1987)

Capping brood	5.55
Early cell cleaning	7.55
Tending brood	9.83
Queen tending	11.1
Patrolling	12.9
Receiving nectar	13.05
Cleaning debris from hive	13.3
Resting/heat shielding	14.15
Late cell cleaning	14.17
First orientation flight	14.17
Comb building	15.5
Handling pollen	16.3
Ventilating	16.85
Guard duty	18.5
First foraging trip	23.43

Fig. 10.2 A chore chart showing the different activities of casts of honey bees. This is an example of a chore chart done by elementary students at Mission Hill Elementary School. The picture is from Classroom Hives (http://media-cache-ec0.pinimg.com/236x/89/f8/f3/89f8f37ab95a3cd18c00d-d3ca5c1d395.jpg)

1. Explain to your class that most of the foods we eat (one out of every three bites) are the result of pollination.
2. Show and read the pages from 4H book, *The Buzz About Bees: Honey Bee Biology and Behavior*. Discuss with the class.
3. Tell your students that they are going to explore a world without bees and, in particular, what the food supply would resemble if bees no longer existed.
4. Direct your students to the "Bee-Free Barbeque" area. Explain that they are going to attend a barbecue in the Bee-Free Zone and that hamburgers and hot dogs are on the menu. Have the students take a few minutes to look at the things they can have with their burger or hot dog.
5. Tell your students to take a plate and choose a hamburger or hot dog from the grill. Explain that they can now choose what they will have with their hamburger or hot dog. Remind them that this is the bee-free barbecue and that the foods on the "Plants Pollinated by Bees" list won't be available. These include tomatoes, onions, cucumbers, lettuce, oil for frying potatoes, oranges, lemons, limes, mustard seed, cacao bean used in making chocolate, vanilla, almonds, watermelon, and apples.
6. Have your students select the items they want with their burger or hot dog. Then have them check the "Plants Pollinated by Bees" list to see what they have to put back.
7. After they've eliminated the bee-pollinated items from their plates, have them describe the meal that would remain.

Students are often amazed to realize that 1 out of every 3 bites of food that we eat is either directly or indirectly dependent on bee pollination. This includes fruit, many nuts (such as almonds), vegetables, oil-seed crops, herbs and spices. In addition, cows graze on clover and alfalfa that are dependent on pollination by bees, so the Bee-Free Barbecue game could potentially also remove hamburger meat, as a final, dramatic impression to leave on students.

Plants Pollinated by Bees (For Use with Pollination Game)
Tomatoes
Onions
Cucumbers
Lettuce
Oil for frying potatoes
Oranges
Lemons
Limes
Mustard seed
Cacao bean (used in making chocolate)
Vanilla
Almonds
Watermelon
Apples

Beyond fruits and vegetables, bees make honey, which is a supersaturated solution. There is too much sugar dissolved in the water than is possible by the laws of physics (over 70% sugar and less than 20% water content). To make honey, a forager bee first sucks up nectar from a flower using its proboscis, a tongue-like set of mouthparts that acts like a straw. Bees store this nectar in a sac organ called a crop, located between the mouth and the stomach. Bees then regurgitate and swallow this nectar, over and over, adding enzymes to break down the sucrose into simple sugars in a process called inversion (Wilson-Rich 2014). Once back in the hive, bees regurgitate a droplet that is taken up by a house bee who also regurgitates and swallows this repeatedly. Then the "honey" is dried by fanning behavior to evaporate the water. Over time, the sugar forms a lattice, crystalline structure, and separates from water. Young students can use this system as a model for designing and creating their own water purification system, such as the following lesson, also from Ms. Williams classroom (see Fig. 10.3).

Materials: You'll need sugar, water, food coloring, a bowl, and a sunny spot.

- Put 20 teaspoons of water into a bowl.
- Stir in 4 tsp sugar.
- Add a drop or two or three of food coloring. Stir.
- Leave the bowl in a sunny spot for a few days.

Fig. 10.3 Students at Mission Hill Elementary School practicing crystallization

- As a control, also leave a bowl of water without sugar in it to compare.
- Check the bowl every day and observe what happens. What happens to the water? What happens to the sugar? Stir the bowl and make observations.
- Optional ... taste test the sugar crystals after the water has evaporated.

Students are often amazed that bees are able to make honey by adding enzymes to nectar in combination with fanning behavior to evaporate the water. They are also surprised to find out that to make a pound of honey, it takes about 556 forager bees visiting two million flowers (National Honey Board 2020)! Therefore, making honey really is a collaborative process.

A good marriage of art, music and STEM education, is through Kinesthetic Activities and games (http://www.classroomhives.org/?page_id=182). One game is the " Flight of the Bumblebee Clean-up/Hive Game". It begins by listening to the video (http://www.youtube.com/watch?v=6QV1RGMLUKE&noredirect=1) and then feeling what it's like to be a bee. Ms. Williams asks the students, "How does this song make you feel?" Example answers include: calm, energetic, crazy, sleepy, thoughtful, like dancing, like running. "Does it want to make you slow down or speed up? What about the music makes you feel that way (tempo, instruments used, similar sharp notes, etc.)?" After listening to and discussing the song as suggested, Ms. Williams then uses it as background music to quickly clean up after a craft, transition to a new activity, or play musical chairs. For Middle School students, while listening to the song they can act out the honey bee round and waggle dances.

The following testimonial provides deep insight into the experiences of Ms. Williams with her classroom beehive. Notice how she takes time to observe the bees personally, and builds an internal sense of inspiration, wonder, and awe that she then transfers and translates to her students. Consider how Ms. Williams leverages bees as a teaching tool to engage her students with an authentic level of excitement that may be able to transcend traditional textbooks and lectures alone. According to Ms. Williams, 3rd and 4th grade teacher at the Mission Hill Elementary School:

> The morning is quiet. It's a warm, sunny, spring day and I am the first on my floor to arrive at school. I put my things down and go to check the hive. As the warm spring weather comes I have to be more diligent about checking in on 1000+ plus 'assistant teachers.' (I'll explain that in a moment.) They honeybees are busy – cleaning cells, leaving to look for pollen and nectar, returning from foraging flights, feeding the brood, looking after the queen. Yes! The Queen in all her magnificence is doing the single most important job she has – laying eggs. I stand and watch in total awe and amazement. As I watch my mind races with questions: 'How old is the queen again? How many eggs did she lay this week? How many drones are in the hive now? How much honey has been made so far? I wonder which student will be the most fascinated by them today?' I look up at the clock and a whole half an hour has passed! I quickly do a water check to make sure the bees are hydrated, and then get back to the work of getting ready for the school day. I had only intended to look at the bees for five minutes or so, but this is the power they hold.
>
> The same goes for students. The honeybees captivate them. In the 13+ years I have had the hive in my classroom I have seen the amazing way that the bees become more than insects in our classroom. The bees become 'teachers' in ways I didn't expect when I first had it installed. They become the teacher who sparks a student's scientific curiosity. They become the teacher who provides challenging math problems. They become the "teacher" who draws out the inner artist in children. They become the teacher who inspires literary greatness, as students write stories, poems, newsletters and songs about them. They become the teacher that illuminates history as the importance of honeybees throughout history is revealed. They become the teacher that calms and soothes students, when anger, frustration, confusion, sadness and fatigue are trying to get the best of them. Having the hive is like having an assistant teacher – another place where students can go to learn, be inquisitive, be inspired, rise to a challenge and feel connected to our classroom community.
>
> One example of this was student M, who was new to our school. He came as a third grader and had some different needs. His hands were physically altered – he only had fingers up to his knuckles. His eyes were not fully straight and he had a hard time seeing. His speech was hindered in a few ways, which made speaking and understanding him difficult. He was smaller in stature than all the other students in the classroom. Another special thing about M. was that he was always smiling – always happy. M. loved the honey bees and took every moment he could to look at them. When he went over to the bees he had a permanent smile. He pointed, he talked to himself, asked questions of us and shared with other children his discoveries. The honeybees were his way 'into' our community. The other students were excited about the things he found. He could always find the queen. This was like his super-

hero power! Other students would go get him and ask him to find her for them. Writing was difficult for him and he avoided it as much as he could, but when it came to the bees he rose to the challenge! He enjoyed drawing pictures of the bees, and never complained when we had him add sentences about what he saw. The bees were the focus of our natural science theme that year and by the end of that three-month period, M. was an expert – sharing all he knew and working with a group of two other boys who created a 3D model of the life cycle of a bee. They used clay and other materials that were difficult for M. to work with, but he dove right in and made important contributions to the project. Most importantly, in the end he was proud of what he had accomplished and his team members saw him as a valued member of the team and our community. The honeybees transformed M. He came in shy, unsure and in the eyes of the other students, 'different.' Through his love of the honeybees he became outspoken, confident, and fully included into our community in ways that would have taken me much longer and may not have been as holistically successful had I relied solely on my 'teacher strategies.'

M.'s needs were unique, but his experience with the honeybees is not. Every year students, and adults alike, make similar connections with the bees. It is a classroom 'project' that has turned into a school-wide treasure.

10.1.1.2 Case Study: Fenway High School

Honey bees in an observation hive can be an effective tool to integrate art and science and give students a creative and engaging way to test their knowledge. Fenway High School has had an observation hive (Fig. 10.4) since 2009 and it is used by Ms. Benadette Manning's mathematics classes. She used a digital microscope projected onto a screen in the classroom in order to visualize the inside of the hive. Students are able to have whole class conversations around their observations and questions using the beehive as a model system, in real time.

Fig. 10.4 The observation hive in the Fenway High School

Patterns in both art and nature can be used to teach students about geometry. The famous graphic artist M.C. Escher's artwork has strong mathematical components including tessellations, using geometric shapes that tile together without any gaps or overlaps. From nature, we can use honeycomb as a tool to get students thinking about different geometric shapes and tessellations. In Ms. Manning's math classes, students practiced calculating volumes, areas, and perimeters of different geometric shapes to determine why bees use hexagons. Students also learned about volume and surface area, using the hexagonal honeycomb shape as a case study. She asks her students, "Why do honey bees use these iso-lateral, 6-sided figures as a tessellated pattern?" Ms. Manning goes through computations of surface area to volume ratios for triangles, squares, rectangles, pentagons, and hexagons. Using these tools of geometry, students come to realize that hexagons have the greatest storage area (volume) to surface area ratio, allowing bees the greatest efficiency between the material resources needed to build comb and the storage reward.

The following insight brings us into the world of what one teenage student in 10th grade experiences and feels when using Ms. Manning's classroom beehive as a learning tool:

> I really enjoyed geometry overall, but a concept I enjoyed out of every topic was learning about bees. I enjoyed learning about bees because I learned why they were important to geometry and my everyday life. For example, I learned that bees connect with geometry because of the shape of their hives. They chose a hexagon because it had way more space than any other figure. Since honey is important to the life of bees they want a maximum amount of honey to store in their hives.

In addition, students can learn about the geometric angles in a worker's bee Waggle Dance (Fig. 10.5). A good video about information on the honey bee waggle dance and experiments to study the waggle dance is, "The Waggle Dance of the Honeybee" by Georgia Tech College of Computing (https://www.youtube.com/watch?v=bFDGPgXtK-U).

Honey bees can be used as a model system to teach NGSS Physical Sciences because they are able to detect odors, tastes, and visual cues (including ultraviolet light). These characteristics enable students to study the detection and responses to different cues. Menzel and Blakers (1975) demonstrated that honey bees have trichromatic vision: UV (350 nm), blue (440 nm), and green (540 nm) photoreceptors. This can be compared to human vision; blue, green and red cones. Humans can see the color red, but bees cannot. Students can design experiments with different colored lights (red, white, and near-UV) and test their hypotheses relating to recognition systems.

Karl von Frisch (1950) first demonstrated that honey bees had associative learning between a sugar-water reward and color cue. Students can design and build artificial flowers of different colors with or without rewards (sugar water and/or pollen) and place them outside to train bees to feed from then. The students can then design an experiment where they either change the orientation (placement) or reward of the flowers to determine if the bees change their foraging behavior (as done with bumble bees by Konzamann and Lanau 2014). At the University of

Fig. 10.5 The geometry of the honey bee waggle dance (by Paige Mulhern)

Arizona, in Dr. Carla Essenberg's laboratory, they designed and built robotic flowers ("FloBots") to test bee foraging behavior (Essenberg 2015).

Patterns can be found in both art and nature. The Fibonacci sequence was written about by Leonardo Fibonacci in his book Liber Abaci (1202). He wrote about this sequence and its relationship to phi (1.618034), also known as the golden ratio (for the English translation, see Sigler 2012). However, Indian philosophers are credited as first noticing a repeating, numerical pattern, within the context of intonation patterns in Sanskrit oral conversation. Start with 0 and 1. Then, add these together to get the next number. 1 + 1 = 2, 2 + 1 = 3, 3 + 2 = 5, 5 + 3 = 8, and so on. This basic pattern forms the infinite sequence of 0, 1, 1, 2, 3, 5, 8, 13, 21, 34, 55, 89….

This pattern is seen in many works of art (Fig. 10.6) including the Mona Lisa (video: https://www.youtube.com/watch?v=jxKYFBtdsqU) and the Vetruvian Man by Leonardo Da Vinci (Belmonte and Pappas n.d.). Art can be created by using tiles made up of squares whose side lengths are successive Fibonacci numbers. The Fibonacci Golden rectangle is found in architecture such as the Greek Parthenon's facade. Fibonacci identified this sequence in many examples in the natural world, especially with regard to spirals found in plants and shells. Plants provide the most obvious examples, including whorled leaf arrangements in trees, seed location in sunflowers, blossom orientation in pineapples, and leaf placement in artichokes. One of the few non-plant examples of the Fibonacci sequence in nature is in the bee. Honey bee DNA structure reflects this patterning in their length and width dimensions and ratio. Genetic relatedness in bees is notable for haplodiploidy, whereby males (drones) are haploid (1 set of chromosomes) and females are diploid (2 sets).

Fig. 10.6 Fibonacci art piece, by Jessica Lindsay

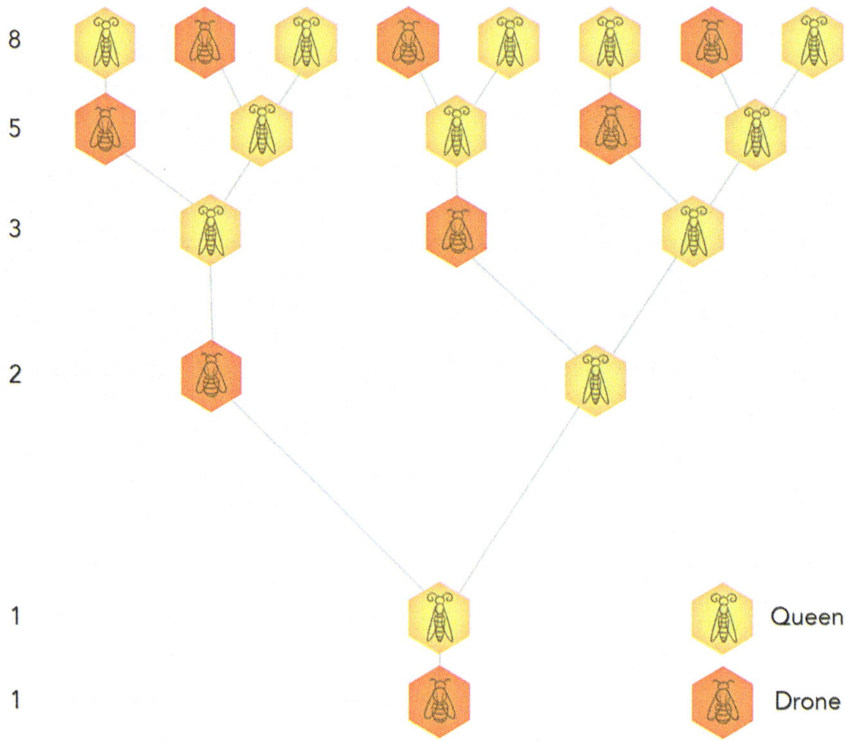

Fig. 10.7 The relationship between honey bee genetics and Fibonacci numbers. Drones (males) are haploid and Queens (female) are diploid. (Figure created by Jessica Lindsay)

Drones arise from unfertilized eggs, and so they have no fathers (however, they do have grandfathers). The family history of drones reflects the Fibonacci sequence: 1 parent, 2 grandparents, 3 great grandparents, 5 great great grandparents… (Fig. 10.7).

In addition to geometry, honey bees can be used as a platform for studying evolution. Flowers and their animal pollinators have co-evolved together. There are an estimated 200,000 species of animal pollinators, with most of them being insects. Pollination involves the transfer of pollen produced by the anther (the male part of the flower) to the stigma (the female part of the flower). Flowers that are insect pollinated attract pollinators with visual cues (including UV), chemical cues, and nectar reward. Bees are well adapted for pollination with adaptations such as eyes that can see UV light, antennae that enable a keen sense of smell, hairy bodies to pick up pollen, and modified mouthparts to suck up nectar. Both bees and flowers benefit from this interaction. Therefore, this relationship between bees and plants is a form of mutualism. Students can observe and draw the adaptations of different types of flowers and insect pollinators that enable their interactions.

Fig. 10.8 Honey bees with the mite *Varroa destructor*. (**A**) A size comparison of a honey bee with a *Varroa* mite. (**B**) A *Varroa* mite attached to a honey bee. (Photo credit: Andrew Bradley)

One of the biggest present threats to honey bee health is the parasitic mite *Varroa destructor. Varroa* is an obligate parasite and they have two life stages: reproductive and phoretic stages. During the *Varroa* phoretic stage, they puncture the honey bees' body to suck their hemolymph (Fig. 10.8A, B). During the *Varroa* mite reproductive phase, they mate on pupae. For a good summary of the *Varroa* life cycle and its impact on honey bees, teachers can show this brief animation (https://www.youtube.com/watch?v=h-wDgd5yURo). Before jumping species to the western honey bee, *Varroa* was a long time parasite of the Asian honeybee, *Apis cerana*. Because of such a long history, both species have coevolved together in an evolutionary arms race (Rosenkranz et al. 2009).

The Asian honeybee, *A. cerana* has evolved defenses against *Varroa* by its grooming ability to detect and eliminate the mite on their fellow workers (Oldroyd 1999). Grooming behavior refers to honeybees removing and disposing of mites on themselves and other workers. Hygienic behavior is the uncapping and removal of dead, diseased, or parasitized brood (Rosenkranz et al. 2009). It is important to note that while these mite defensive behaviors have been noticed in some European honey bee (now predominant in the New World, as well; *A. mellifera*) colonies, they occur at a limited extent when compared to the sister species *A. cerana* (Boecking and Spivak 1999). Restricting the mites to drones and drone brood cause the mites to be highly limited in their ability to disrupt the colony, and thus *Varroa* and *A. cerana* could coexist. However, as a result of human globalization, *Varroa* was introduced to *A. mellifera* colonies.

When *Varroa* jumped host species to the western honeybee, the western honeybee did not have the same defenses as the Asian honey bee, and the results were catastrophic (Oldroyd 1999). The effect of the mite's species jump coupled with the lack of any parasitic influence on the genetic lineage of *A. mellifera* has left the species mostly defenseless against the *Varroa* parasite infection. The mite becomes extremely detrimental to the health of the hive because it is a vector for multiple viruses including Kashmir bee virus, Sacbrood virus, Acute bee paralysis virus, Israeli acute paralysis virus, and deformed wing virus (Oldroyd 1999; Rosenkranz et al. 2009). If left untreated, *Varroa* infections often contribute to the death or absconding of an entire colony (Oldroyd 1999). For a good summary of the *Varroa* life cycle and its impact on honey bees, teachers can show this brief animation (https://www.youtube.com/watch?v=h-wDgd5yURo).

The relationship between *A. cerana, A. mellifera*, and *Varroa,* provides teachers with a very real and relevant example of many globally important ecological concepts, including vector biology, co-evolution, and epidemiology. Additionally, this study system can be scaled up to engage students in conversation and thoughts relating to globalization in our modern world, and the all-around effects of human behavior on the environment. *Varroa* was originally introduced to *A. mellifera* due to human movement and transportation of bees. This species jump from *A. cerana* to *A. mellifera* is one of the many potential consequences of introducing non-native species to new ecosystems. Students are exposed to the many consequences of globalization as the world continues to become increasingly interconnected.

Since *Varroa* is such an extreme detriment to honeybee health, effective treatments for *Varroa* has become a keynote area of study for scientists and beekeepers alike. Because of the colony destruction and overall economic impact of depleting honeybee colonies, many commercial and hobbyist beekeepers alike have turned to the use of miticides like Apistan and Checkmite+ to control mite populations in colonies. Given the short life cycle and inbreeding of *Varroa*, resistance to these miticides occurs rapidly (Hubert et al. 2013). In what is referred to as a bottleneck effect, small populations of mutated mites spawn a larger population of resistant family members. With reproductive success of *Varroa* on *A. mellifera* so high, killing off non-resistant populations favors the establishment of a resistant population.

Varroa resistance to miticides presents teachers with another real world example of an evolutionary concept: the bottleneck effect. The bottleneck effect is an important concept in population ecology because it can help students understand macroevolutionary concepts, including speciation and extinction, can occur in relatively short periods of time. The short life-span and reproductive success of *Varroa* makes their bottlenecking due to miticides particularly interesting for students. The relationship between *Varroa* and miticides can be extrapolated to human health by examining bacteria and their developing resistance to antibiotics. These types of concerns will continue to arise if humans continue to rely primarily on extermination methods like pesticides, miticides, and antibiotics to kill unwanted populations of pests, mites, and diseases.

Of course, with such a widespread issue, many different non-chemical maintenance methods have been tested and suggested. For example, some beekeepers have tried wire-screened, bottom boards to eliminate mites that have fallen from the colonies. Others have tried pouring powdered sugar dusting to induce cleansing behavior of worker bees. In an effort to find a sustainable solution to the *Varroa* mite problem in honey bees, students in Dr. Marla Spivak's bee laboratory at the University of Minnesota (http://www.beelab.umn.edu/) selectively bred honey bees that have exhibited hygienic behavior, which includes grooming off *Varroa* mites. Bees with hygienic behavior maintain lower mite populations in their colonies (Spivak and Reuter 2001; also see https://www.youtube.com/watch?v=W_0FPF1Smwk).

Some scientists have gone even farther and begun the process of isolating genes responsible for hygienic behavior on both *A. cerana* and *A. mellifera.* For example, in an attempt to compare and contrast the two species' activity in response to a

Varroa infection, one group studied differential gene expression of *A. cerana* and *A. mellifera* in the presence of the mite (Zhang et al. 2010). Another group did a similar study on two genetic stocks of *A. mellifera*, which differ in susceptibility to *Varroa* infection. Both studies showed significant differences in gene patterns of expression, signifying a genetic predisposition for hygienic behavior.

Blending science with other fields of study and inquiry are common at many schools that integrate subjects for middle-aged children. Ms. Jacqueline Beaupré brought her experiences in beekeeping into the classroom, where she consistently turns her students' attention to the bees for a living application of teaching concepts relating to STEAM. Notice how Ms. Beaupré's sense of wonder seems to translate to her students, by building excitement for learning in an authentic and inspiring way. Her passion for the bees and these topics help tie concepts together in a way that goes beyond traditional teaching methods alone. According to Ms. Beaupre:

> Bees are fascinating to watch! I think every curious kid goes through a 'bug phase' and they all slip back into it when looking at our (classroom) observation hive. There is no 'free time' in (our) high school classes, but I notice students taking a moment to observe the bees when transitioning between activities or before the bell rings. They usually come away with a question about the myriad of hive activities they see- workers dancing to communicate, cleaning, building wax, making honey, and feeding brood, as well as the queen laying eggs. Very often, they notice important hive activities before I do!
>
> Honeybees are the perfect animals for my Integrated Science classroom as I can connect them to many aspects of physics, chemistry, and biology in our curriculum. Bees take nearly direct routes to foraging sites (distance vs. displacement, Traveling Salesman Problem) and can easily attract pollen to their hairy bodies (electrostatic force, charges). They eat pollen and nectar, giving them the nutrients they need to excrete royal jelly for feeding their young and beeswax for building their home (biomolecules- protein, carbohydrates, lipids). They carefully thermoregulate their hive, surviving winter by eating honey and vibrating muscles to keep their colony warm (transformation of energy, heat transfer). They can see ultraviolet light in addition to colors (waves, electromagnetic spectrum). They have mutually beneficial partnerships with the flowers they pollinate, as well as tenuous ones with many gut microbes, viruses, and hive pests (botany, evolution, ecological relationships).
>
> As my own knowledge about science and bees increases, I continue to find connections to share with students. And although I could include these anecdotal examples in my lessons without a classroom hive, its physical presence produces a much more meaningful experience. When students can actually taste honey from our hive, it makes a lesson about biopolymers and disaccharides far more memorable. In this way, I feel our students' learning is truly enhanced by our classroom observation hive.

10.1.2 CASE 2: Langstroth Hive

In 1852, L.L. Langstroth patented his design for the stacked-box beehives with removable, hanging frames that are commonly used today. With traditional Langstroth hives, disease monitoring is crucial to colony health. The removable parts allow for full inspection and visualization of every part of the inner beehive. This innovation outpaced that of the prior beehive designs without removable parts, typically of the basket hive design, called, 'skeps.' Few other designs for beehives have since been created, including 'top bar' beehives. The ease of beekeeper inspections with Langstroth beehives makes these the leading style for outdoor beekeeping throughout developed countries.

Myriad pests and pathogens are prevalent throughout honeybee hives worldwide, and beekeepers are on the lookout for each during routine inspections of Langstroth beehives. One modern plague of honey bees is the fungal infection, *Nosema ceranae*, which is associated with decreased honey production, precocious foraging (younger bees begin foraging at a younger age than normal), decreased lifespan, and colony death (Botías et al. 2013; Goblirsch et al. 2013; Higes et al. 2009). *N. ceranae* is considered to be one of the most prevalent honey bee pathogens (Traver and Fell 2011). *Varroa destructor* are mites that feed on the hemolymph of bees and are vectors of several viruses, like deformed wing virus (DWV). If mites feed on developing bee pupae, the developed adult's immune system can be suppressed (Yang and Cox-Foster 2007). Dainat et al. (2012) found in the winter, *Varroa destructor* and DWV were predictive markers for honey bee colony death. It is important to estimate mite concentrations for each colony, due to their destructive nature on the health of the colony.

Modern solutions for updating Langstroth's designs are on the horizon. These provide students and researchers with ample opportunities for engaging STEM education and learning, with special emphasis on technology and engineering. Smart hives connect the Internet of Things to beehives, by installing a computer data collection system inside the hive. Data are collected in real-time, transmitted through the cloud, and reported back to classrooms through an online portal. Sensors throughout the beehive monitor and collect data relating to temperature, humidity, weight, sound, and other metrics, including live video streaming. The Best Bees Company deployed the world's first SmartHive at the Museum of Science, Boston, working closely with their education and curatorial staff to integrate STEM learning opportunities alongside their live beehive exhibit. Using this technological and engineering data collection tool, students can now test hypotheses relating to beehive thermoregulation, behavioral ecology, environmental biology, and more.

A similar advancement in technology and engineering powering modern beekeeping is happening in the United Kingdom. Arnia hive monitoring systems allow beekeepers and students to track the progress of their hives using cloud-based streaming data. The Arnia monitors can detect external weather conditions, in-hive temperature and humidity, hive weight, and even interpret the sounds of the bees to assess colony behavior and health (Figs. 10.9, 10.10, and 10.11). Arnia Remote Hive Monitoring (http://www.arnia.co.uk/) has partnered with schools and businesses in a Citizen Science program called Build the Buzz, a bee listening project. The data from these monitors can be viewed from any internet-enabled device. These data allow participants the ability to view and compare colony development and activity from different colonies that are affected by pests (e.g., wax moths),

Fig. 10.9 This shows the user interface from the monitored hives with current sensor readings and weather conditions (from Arnia Hive Monitoring Systems)

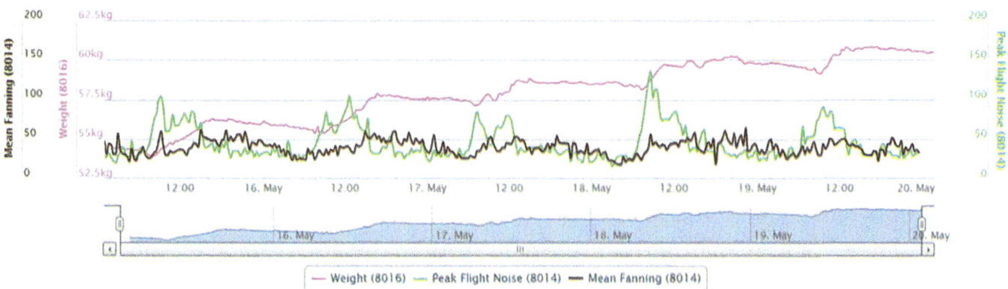

Fig. 10.10 This graph from the Arnia Hive Sensors shows the nectar flow and processing in a hive. The weight (pink line) increases during the day as nectar is collected by foraging bees (green line shows corresponding daytime peaks in flying activity). Weight decreases slightly at night as the nectar is processed by bees fanning (brown line shows increase in nighttime fanning)

parasites (e.g, *Varroa* mites) and pathogens (e.g., bacterial and fungal infections such as foulbrood and *Nosema*, respectively).

10.1.2.1 Case Study: Northeastern University Co-op Students at The Best Bees Company

To quantify *Nosema* levels, spores can be visually calculated using the technique from Reuter et al. (University of Minnesota Instructional poster #167, https://www.beelab.umn.edu/sites/beelab.umn.edu/files/cfans_asset_317468.pdf). However, this

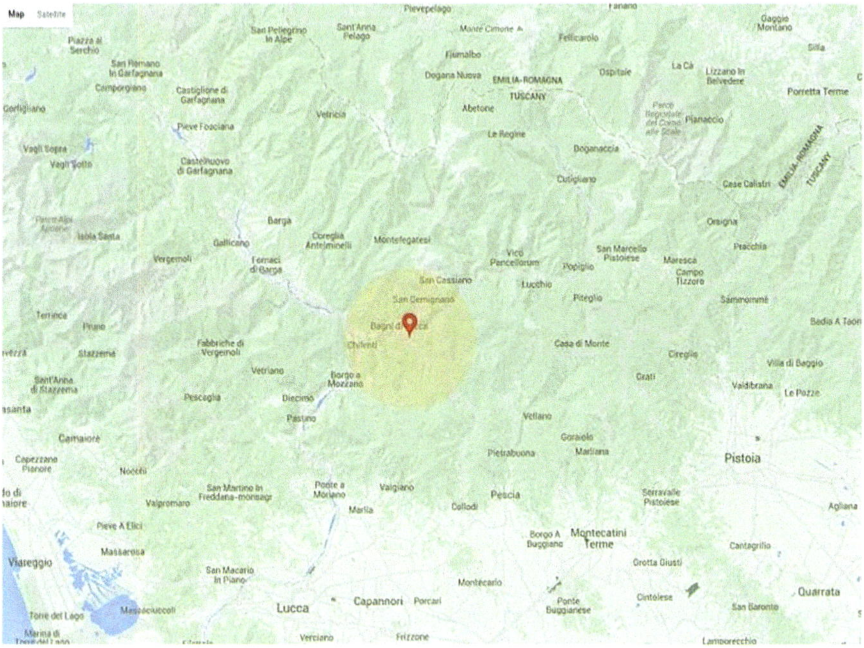

Fig. 10.11 With the Arnia In-hive monitors, you can locate your hive and its forage map area

technique cannot distinguish between the two different species of *Nosema* (*N. apis and N. ceranae*). Northeastern University Co-op student interns at The Best Bees Company have collected and analyzed data on *Nosema* spore counts and *Varroa* mite counts since 2013. These college students found a statistical difference between alive and dead beehives in the presence of *Nosema* and *Varroa* (Chi-square, df = 3, Chi-Square = 66.03, $p < 0.0001$; Fig. 10.12). Dead hives had a higher prevalence of having both *Nosema* and *Varroa* compared to alive and healthy hives.

In a marriage of science and art, one of Northeastern University's Digital Art and Interactive Media students, Jessica Lindsay is working with The Best Bees Company as the Webmaster and Graphic Design Assistant (she created Figs. 10.1, 10.6, and 10.7). In addition, at The Best Bees Company, Jessica has painted many custom beehives, designed various marketing materials, translated research data into info-graphics, and created webpages for the company. She is currently collaborating with another artist at The Best Bees Company, Paige Mulhern, a former MassART student (who made Figs. 10.5 and 10.12), hand-paints illustrations of flowers that accompany data reports for the honeyDNA program (https://bestbees.com/product/honeydna/, https://www.ted.com/talks/noah_wilson_rich_how_you_can_help_save_the_bees_one_hive_at_a_time, and https://www.nationalgeographic.com/magazine/2018/02/a-dollop-of-sweet-science/). HoneyDNA is a publicly available genomics service that sequences the DNA of all pollen spores in honey samples. Paige translates the data reports into art, allowing for improved scientific

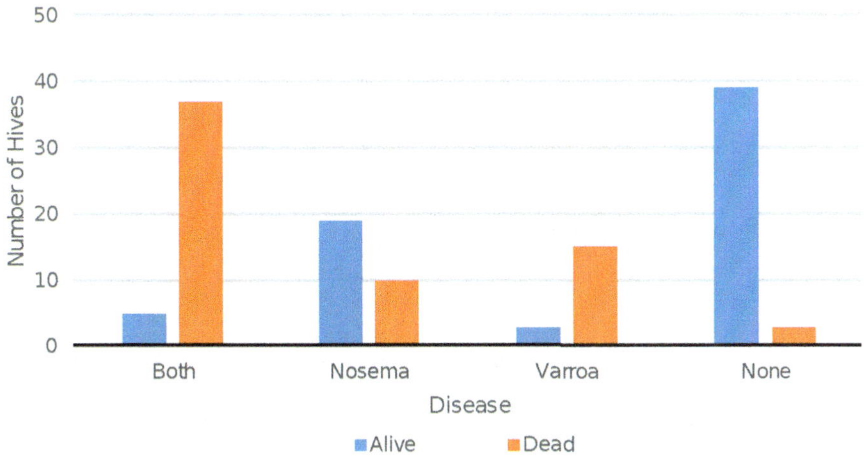

Fig. 10.12 Mean number of alive (blue) or dead (orange) colonies that had both *Nosema spp*. and *Varroa*, just *Nosema*, just *Varroa*, or neither. Hives were located in the greater Boston area

communications with the general public through infographics. Students and learners of all ages can use these data responsively, by selectively planting seeds that the honeyDNA genomics technology results show are exactly what bees forage upon.

The following is a description of the honeyDNA genomics technology program.

> Knowing exactly what type of honey your bees produce adds value, both on the markets and in the experience. You can identify the floral source of your honey with our new, state of the art product, honeyDNA. We send you a bar-coded test tube for privacy, along with a stamped return envelope to our laboratory. You fill the test tube with 10 mL – 20 mL of honey. Our research team analyzes the DNA of all the pollen found in your honey. We then compare the pollen genomes to our comprehensive database of known plant families. Within six to eight weeks, we will deliver to you the results in the form of a custom infographic, with hand-painted illustrations of the flower families found in your sample of honey.

We can look to nature for sustainable solutions for human problems in a field of study called biomimicry or biomimetics (Pawlyn 2011). Honey bees can inspire new technology while teaching students about cognitive function and neurobiology. In a joint project with Harvard University and Northeastern University, students are working on developing an Autonomous Flying Microrobots that mimics bees, "Robobees" (http://wyss.harvard.edu/viewpage/457). Robobees have three components, each scaling levels of biological organization: brain, body, and colony. The brain component has "smart" sensors that mimic the bee antennae and eyes that are connected to control electronics so they can sense and respond to stimuli in their environment. The colony component is working on coordinating multiple independent Robobees so they can work together.

Geographic Information System (GIS) technology is a useful tool to study honey bees and their habitat (reviewed in Rogers and Staub 2013). Naug (2009) used GIS to study honey bee losses for each state in the United States. States with the lowest

Fig. 10.13 Honey production in urban, suburban, and rural areas in Massachusetts from 2010–2013. (created by Paige Mulhern)

amount of open space (not developed) had the lowest levels of honey production. Giannini et al. (2012) used GIS to study the impacts of predicted climate change on 10 Brazilian native bee species. Urban areas in the United Kingdom were shown to have a higher species richness of flower visiting insects compared to farmland (Baldock et al. 2015). Senapathi et al. (2015) also found a high species richness in bees and wasps in urban areas compared to farmland in England. But Rathcke and Jules (1994) found that flowering plants in fragmented habitats have reduced visits by insect pollinators.

A surprisingly large number of bee species (200) have been found in New York City (Great Pollinator Project), including six phylogenetic families of bees including 30 exotic bee species. This may reflect the fact that most urban areas also support a higher plant species richness (Grimm et al. 2008). In Boston and Cambridge, Massachusetts, USA there has been high recorded survival and productivity of urban managed honey bee hives (from 2010–2013; Fig. 10.12). These urban hives had a higher overwinter survival and produced more honey compared to rural hives (Wilson-Rich 2014; Fig. 10.13). Some possible hypotheses for the increased fitness and productivity of urban bees are urban heat-island effect, less pesticide use in the city, a varied diet of flowers, and less energy exposure for foraging (Wilson-Rich 2014).

10.1.3 CASE 3: Citizen Science Projects (with Wild Bees)

If neither observation nor Langstroth beehives are available, then there are still more learning opportunities that exist with urban bees. A helpful website is PollinatorLIVE: a distance learning adventure (http://pollinatorlive.pwnet.org/index.php), which includes teacher resources, webcasts (in English and Spanish), and links to several citizen science projects. This website includes lesson plans and how they correlate with national science standards. Multiple Citizen Science Projects focus on pollinators, such as the Great Sunflower Project (https://www.greatsunflower.org/) and the Bees' Needs (http://beesneeds.colorado.edu). The Great Sunflower Project has a Google Earth map that shows the average pollinator visits (from 2008–2012) for Lemon Queen Sunflowers (https://www.greatsunflower.org/Map) across the United States. The results indicate that urban areas tend to have the least pollinator service. The Goldenrod Challenge (http://www.discoverlife.org/goldenrod/) is using photography and a systems-approach to learn about the species interactions at goldenrods. Two Citizen Science projects focus on the impacts of climate change on phenology: (1) Project BudBurst (http://budburst.org/) and, (2) the USA-National Phenology Network (NPN; https://www.usanpn.org/). In conjunction with Project BudBurst, students can learn about the coevolution of plants and pollinators. The USA-NPN has an on-line Nature's Notebook to track seasonal changes in the timing of insect emergence, flowering, and bird migration in relation to climate. This data has been used by scientists at Yale University to determine that the timing of budburst has changed over time (1982–2012) in the Eastern region of the U.S. and that this is a direct response to warming trends (Yue et al. 2015).

10.1.4 CASE 4: Pollinator Gardens and Insect/Bee Hotels

Build it and they shall come. The U.S. Fish and Wildlife Program has a Schoolyard Habitat Program (https://www.fws.gov/CNO/conservation/Schoolyard.html) with helpful links and a useful document on how to design, build, and implement a schoolyard habitat (https://www.fws.gov/cno/pdf/HabitatGuideColor.pdf). A good non-GMO pollinator seed mix called "Save the Bees" is available from Botanical Interests® (Fig. 10.14a–d) (https://www.botanicalinterests.com/products/view/7020/Flower-Mix-Save-the-Bees-Seeds-LG). This seed packet needs to be planted in full sun and these flowers do attract a diversity of bees (Fig. 10.14B–D). Once the flowers have bloomed, each student can calculate the insect pollinator service by counting the number of pollinators attracted to a single flower in a 15 minute time span. The class can then add up each student's pollinator service data to get a total pollinator service, which is divided by the number of student observations to calculate an average pollinator service quotient for your garden. Make sure to record controlled variables, such as the species of pollinator (you can take pictures to later identify species), temperature, weather conditions, and time of day. In addition, this can be repeated over the course of a season or more to view changes in pollinator service over time. These data can then be plotted to visualize trends in the results.

Photography is a successful way to engage students with nature. The students can design art projects in the garden, such as taking sequential pictures from the same spot of the garden over time to track the phenology (timing of flowers blooming, insect emergence, etc.). Photography is useful in identifying different species of

Fig. 10.14 (**A–D**) Flowers grown from the seed packet "Save the Bees" from Botanical Interests®. (**B**) A rusty patched bumble bee on a giant sunflower leaf in the garden. (**C**) A honey bee pollinating a star flower. (**D**) A bumble bee and a metallic green bee ("sweat bee") pollinating a sunflower. (Photo credit: Kristian Demary)

plants and pollinators. Mobile app technology, such as FlowerPedia 2.0, Botany Buddy, and Audubon Wildflowers (https://dengarden.com/gardening/Free-Smartphone-Apps-for-Plant-Lovers) are useful to help identify flowers.

10.1.4.1 Case Study: Massachusetts College of Art and Design

In addition to planting flowers, students can design and build habitat (or nesting sites) for urban insects and bees. Mason bees are solitary bees that build a nest inside tunnels left behind by tree boring insects. Mason bees cannot dig the tunnels themselves, therefore they must look for an abandoned one. Humans can supply them with all the tunnels they need by providing wood with drilled holes or hollow reeds. A good reference on how to build insect hotels can be found at: http://www.foxleas.com/make-a-bee-hotel.asp.

The students in Dr. Kristian Demary's Sustainability Science course at Massachusetts College of Art and Design (MassART), along with resources from The Best Bees Company and the Urban Beekeeping Laboratory and Bee Sanctuary

Fig. 10.15 The bee hotel was built by Massachusetts College of Art and Design students on April 29, 2016. (Photo credit: Kristian Demary)

(www.urbanbeelab.org) created habitat for solitary bees with a bee hotel out of cardboard boxes that included wood with holes drilled in it along with hollow reeds (Fig. 10.15). In addition, Dr. Demary and her students upcycled unwanted products and waste into useful habitat for a diversity of insects and other small animals that is on display in the MassART courtyard (Fig. 10.16). They used five wood pallets as a base for the insect hotel. The students collected recyclable and natural materials over the course of the semester. They planted flower seeds in glass bottles and cans to include a nectar source for the wild bees.

Dr. Demary's students were surprised to know that they could upcycle materials that would normally be either thrown in the trash (old bricks, wood pallets, and fabric and flooring samples) or recycled (glass and plastic bottles, coffee cups, cardboard boxes, egg cartons, aluminum cans, and cork) along with natural items (reeds and sticks), as building material to make useful habitat for insects and other small animals to use. After two months, at least three mason bees made nests in the drilled holes (Fig. 10.17). After three months, they found ants, slugs, spiders, pill bugs, and a white millipede in, around, and under the MassART insect hotel (Fig. 10.18A–D). In addition, some of the seeds they had planted germinated and bloomed into flowers (Fig. 10.19). Dr. Demary continues to monitor the insect and bee hotels to determine the best building materials for insect habitat.

Fig. 10.16 The insect hotel designed and built by Massachusetts College of Art and Design students on April 29, 2016. It was made out of recyclable and natural materials. (Photo credit: Kristian Demary)

10.2 Summary

In summary, urban bees are an established and successful platform for STEAM education. STEAM education not only fosters curiosity and creativity, but it can also lead to an interdisciplinary approach to solving real-world problems. Bees are an effective means with which to advance STEAM education for learners of all ages, socioeconomic backgrounds, origins, nationalities, races, and interests.

Fig. 10.17 Students drilled holes in wood for Mason Bees and three were used as nests (the ones with mud). This picture was taken on June 24, 2016. (Photo credit: Kristian Demary)

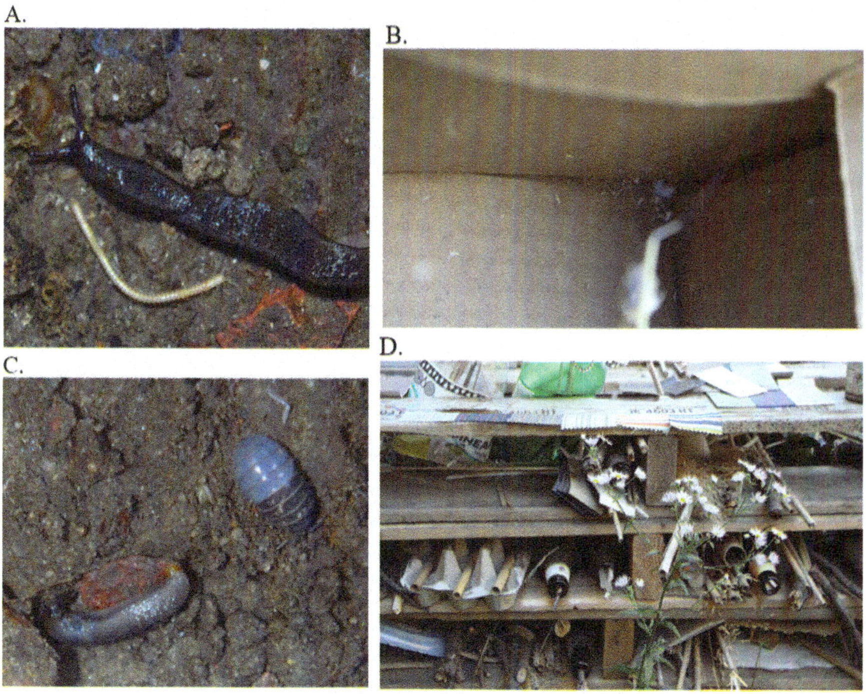

Fig. 10.18 Life found in, around, and under the Bee Hotel and the Insect Hotel three months after they were built. The Insect and Bee Hotels are on display in the courtyard at Massachusetts College of Art and Design. (**A**) A slug and millipede found under a cardboard box of the Bee Hotel. (**B**) A spider in a cardboard box of the Bee Hotel. (**C**) A pill bug (going through molting) and a slug found under a wood board. (**D**) plant life around the insect hotel. (Photo credits: **A** and **C**: Kristian Demary; **B** and **D**: Andrew Bradley)

Fig. 10.19 Three months after the Insect Hotel was built and the seeds were planted, we found flowers in bloom. This is on display in the courtyard at Massachusetts College of Art and Design. (Photo credit: Kristian Demary)

References

Baldock, K. C. R., Goddard, M. A., Hick, D. M., Kunin, W. E., Mitschunas, N., Osgathorpe, L. M., Potts, S. G., Robertson, K. M., Scott, A. V., Stone, G. N., Vaughan, I. P., & Memmott, J. (2015). Where is the UK's pollinator biodiversity? The importance of urban areas for flower-visiting insects. *Proceedings of the Royal Society B, 282*, 20142849.

Belmonte, C., & Pappas, C. (n.d.). *Painting by numbers: The Fibonacci sequence in art*. Retrieved from http://www.scientiareview.org/pdfs/208.pdf

Boecking, O., & Spivak, M. (1999). Behavioral defenses of honeybees against *Varroa jacobsoni*. *Apidologie, 30*, 141–158.

Botías, C., Martín-Hernández, R., Barrios, L., Meana, A., & Higes, M. (2013). Nosema spp. infection and its negative effects on honey bee (Apis mellifera iberiensis) at the colony level. *Veterinary Research, 44*(25) Retrieved from http://www.ncbi.nlm.nih.gov/pmc/articles/PMC3640932/.

Czech, B. (2004). Urbanization as a threat to biodiversity: Trophic theory, economic geography, and implications for conserving land acquisition. In *Proceedings of a Symposium at the Society for Conservation Biology 2004 Annual Meeting*. Retrieved from http://www.nrs.fs.fed.us/pubs/gtr/gtr_nc265/gtr_nc265_008 pdf

Czech, B., Krausman, P. R., & Devers, P. K. (2000). Economic associations among causes of species endangerment in the United States. *BioScience, 50*, 593–601.

Dainat, B., Evans, J. D., Chen, Y. P., Gauthier, L., & Neumann, P. (2012). Predictive markers on honey bee colony collapse. *PLoS One, 7*, e32151. http://www.ncbi.nlm.nih.gov/pmc/articles/PMC3285648/.

Essenberg, C. J. (2015). Flobots: Robotic flowers for bee behaviour experiments. *Journal of Pollination Ecology, 15*(1), 1–5. Retrieved from http://www.ncbi.nlm.nih.gov/pubmed/25722755.

Fisher, M., Miller, D., & Jamison, K. (2014). *The 4H Honey bee youth project book 1 "The Buzz about Bees: Honey bee biology and behavior"*. Retrieved from https://pubs.ext.vt.edu/380/380-070/380-070_pdf.pdf

Giannini, T. C., Acosta, A. L., Garófalo, C. A., Saraiva, A. M., Alves-Dos-Santos, I., & Imperatriz-Fonseca, V. L. (2012). Pollination services at risk: Bee habitats will decrease owing to climate change in Brazil. *Ecological Modelling, 244*(0), 127–131.

Goblirsch, M., Huang, Z. Y., & Spivak, M. (2013). Physiological and Behavioral Changes in Honey Bees (*Apis mellifera*) induced by *Nosema ceranae* infection. *PLoS*, e58165. Retrieved from http://www.ncbi.nlm.nih.gov/pmc/articles/PMC3590174/.

Grimm, N. B., Faeth, S. H., Golubiewski, N. E., Redman, C. L., Wu, J., Bai, X., & Briggs, J. M. (2008). Global change and the ecology of cities. *Science, 319*, 756–760. https://doi.org/10.1126/science.1150195.

Higes, M., Martín-Hernández, R., Garrido-Bailón, E., González-Porto, A. V., García-Palencia, P., Meana, A., del Nozal, M. J., Mayo, R., & Bernal, J. L. (2009). Honeybee colony collapse due to *Nosema ceranae* in professional apiaries. *Environmental Microbiology Reports*. https://doi.org/10.1111/j.1758-2229.2009.00014.x. http://www.ncbi.nlm.nih.gov/pubmed/23765741.

House Resolution 319. (2011). *Expressing the sense of the House of Representatives that adding art and design into Federal programs that target the Science, Technology, Engineering, and Mathematics (STEM) fields encourages innovation and economic growth in the United States*. https://www.govtrack.us/congress/bills/112/hres319/text

Hubert, J., Nesvorna, M., Kamler, M., Kopecky, J., Tyl, J., Titera, D., & Stara, J. (2013, November). Point mutations in the sodium channel gene conferring tau-fluvalinate resistance in *Varroa destructor*. *Pest Management Science*. https://doi.org/10.1002/ps.3679.

Jaganmohan, M., Vailshery, L. S., & Nagendra, H. (2013). Patterns of insect abundance and distribution in urban domestic gardens in Bangalore, India. *Diversity, 5*, 767–778. https://doi.org/10.3390/d5040767.

Jones, E. L., & Leather, S. R. (2012). Invertebrates in urban areas: A review. *European Journal of Entomology, 109*, 463–478.

Klein, A. M., Vaissière, B. E., Cane, J. H., Steffan-Dewenter, I., Cunningham, S. A., Kremen, C., & Tscharntke, T. (2007). Importance of pollinators in changing landscapes for world crops. *Proceedings of Biology Sciences, 274*, 303–313.

Konzamann, S., & Lanau, K. (2014). Divergent rules for pollen and nectar foraging bumblebees-A laboratory study with artificial flowers offering diluted nectar substitute and pollen surrogate. *PLoS ONE, 9*(3), e91900, Retreived from https://doi.org/10.1371/journal.pone.0091900

Maeda, J. (2013). *STEAM: Adding art and design to STEM*. Retreived from http://arcadenw.org/article/steam

Menzel, R., & Blakers, M. (1975). Colour receptors in the bee eye – Morphology and spectral sensitivity. *Journal of Comparative Physiology. A, Neuroethology, Sensory, Neural, and Behavioral Physiology, 108*, 11–13. http://link.springer.com/article/10.1007/BF00625437.

Morse, R., & Calderone, N. (2000). The value of honey bees as pollinators of US crops in 2000. *Bee Culture, 128*, 1–14.

National Honey Board. (2020). *Honey Trivia*. Retreived from http://www.honey.com/newsroom/press-kits/honey-trivia

Naug, D. (2009). Nutritional stress due to habitat loss may explain recent honey bee colony collapses. *Biological Conservation, 142*(10), 2369–2372. https://doi.org/10.1016/j.biocon.2009.04.00.

Next Generation Science Standards. Retreived from http://www.nextgenscience.org/

Oldroyd, B. P. (1999). Coevolution while you wait: *Varroa jacobsini*, a new parasite of western honeybees. *Trends in Ecology & Evolution, 8*, 312–315.

Pawlyn, M. (2011). *Biomimicry in architecture*. London: RIBA Publishing.

Rogers, S. R., & Staub, B. (2013). Standard use of Geographic Information System (GIS) techniques in honey bee research. *Journal of Apicultural Research, 52*(4). https://doi.org/10.3896/IBRA.1.52.4.08.

Rosenkranz, P., Aurneier, P., & Ziegelmann, B. (2009). Biology and control of *Varroa destructor*. *Journal of Invertebrate Pathology*, 96–119.

Senapathi, D., et al. (2015). The impact of over 80 years of land cover changes on bee and wasp pollinator communities in England. *Proceedings of the Royal Society B, 282*, 20150294. https://doi.org/10.1098/rspb.2015.0294.

Sigler, L. E. (2012). *Fibonacci's Liber Abaci: A translation into modern English of Leonardo Pisano's book of calculations*. New York: Springer.

Spivak, M., & Reuter, G. S. (2001). Varroa destructor infestation in untreated honey bee colonies selected for hygienic behavior. *Journal of Economic Entomology, 94*, 326–331.

STEM to STEAM. Retreived from http://stemtosteam.org/

Traver, B. E., & Fell, R. D. (2011). Prevalence and infection intensity of *Nosema* in honey bee (Apis mellifera L.) colonies in Virginia. *Journal of Invertebrate Pathology, 107*, 43–49. http://www.ncbi.nlm.nih.gov/pubmed/21345338.

von Frisch, K. (1950). *Bees; their vision, chemical senses, and language*. Ithaca: Cornell University Press.

Wilson-Rich, N. (2014). *The bee: A natural history*. Princeton: Princeton University Press.

Winston, M. (1987). *The biology of the honey bee*. Cambridge, MA: Harvard University Press.

Yang, X., & Cox-Foster, D. (2007). Effects of parasitization by *Varroa destructor* on survivorship and physiological traits of *Apis mellifera* in correlation with viral incidence and microbial challenge. *Parasitology, 134*(part 3), 405–412.

Yue, X., Unger, N., Keenan, T., Zhang, X., & Vogel, C. S. (2015). Probing the past 30-year phenology trend of U.S. deciduous forests. *Biogeosciences, 12*, 4693–4709. http://www.biogeosciences.net/12/4693/2015/doi:10.5194/bg-12-4693-2015.

Zhang, Y., Liu, X., Zhang, W., & Han, R. (2010). Differential gene expression of the honey bees *Apis mellifera* and *A. cerana* induced by *Varroa destructor* infection. *Journal of Insect Physiology, 56*, 1207–1218.

Index

I. DeCoito et al. (eds.), *Teaching and Learning in Urban Agricultural Community Contexts*, Urban Agriculture,
https://doi.org/10.1007/978-3-030-72888-5

The manufacturer's authorised representative in the EU is Springer Nature Customer Service Centre GmbH, Europaplatz 3, 69115 Heidelberg, Germany. If you have any concerns regarding our products, please contact ProductSafety@springernature.com

Printed and bound by CPI Group (UK) Ltd, Croydon, CR0 4YY
30/09/2026
02225491-0001